Universitext

T0155688

Springer
New York
Berlin
Heidelberg
Barcelona
Hong Kong
London
Milan
Paris
Singapore
Tokyo

Universitext

Editors (North America): S. Axler, F.W. Gehring, and K.A. Ribet

(continued after index)

Olivier Debarre

Higher-Dimensional Algebraic Geometry

With 16 Illustrations

Springer

Olivier Debarre
IRMA–Université Louis Pasteur–CNRS
7, rue René Descartes
67084 Strasbourg Cédex
France
debarre@math.u-strasbg.fr

Mathematics Subject Classification (2000): 14-01, 14Jxx

Library of Congress Cataloging-in-Publication Data
Debarre, Olivier.
    Higher-dimensional algebraic geometry / Olivier Debarre.
        p. cm. — (Universitext)
    Includes bibliographical references and index.

    1. Geometry, Algebraic.   I. Title.   II. Series.
QA564 .D44 2001
516.3′5—dc21   2001020053

Printed on acid-free paper.

Photocomposed copy prepared from the author's LaTeX2e files.

9 8 7 6 5 4 3 2 1

ISBN   978-1-4419-2917-4

Springer-Verlag   New York Berlin Heidelberg
A member of BertelsmannSpringer Science+Business Media GmbH

Preface

This book deals with the classification theory of algebraic varieties. Traditionally, "higher-dimensional" refers to the case of dimension at least 3 (or, more accurately, of any dimension), as opposed to the case of surfaces, where everything is completely understood.

The theory is largely still in progress. However, an amazing quantity of knowledge has accumulated in the past twenty years, through the combined efforts of S. Mori, F. Campana, Y. Kawamata, J. Kollár, Y. Miyaoka, M. Reid, V. Shokurov, and many others. We are now at a stage where understanding the latest developments requires mastering a multitude of complex and subtle notions. One sometimes feels one has to join a private club where people talk freely about, say, the discrepancy of weak Kawamata log terminal pairs or the existence of general elephants.

As a result, the field has remained much more confidential than it should have. Strong students or seasoned mathematicians have been known to flinch before the task of learning the terminology of the field, discouraged by the reputation of technicality surrounding it or simply not knowing where to begin.

My purpose in writing these notes was to provide an easily accessible introduction to the subject. I have therefore not tried to be exhaustive, nor to write a reference book. To begin with, such books already exist (I am thinking of Kollár's book [K1] on rational curves). Secondly, the field is moving ahead rapidly. Thirdly, I am certainly not the expert that such an undertaking would require. I have, on the contrary, selected small parts of the theory and tried to give definitions, proofs, and examples with as many details as possible. This means that the experts will not find anything new here, except maybe some marginal new results on Fermat hypersurfaces (see Exercises 2.5) and Fano varieties with high degree (see Section 5.11).

The material covered in this book falls roughly into three groups.

• The first two chapters are devoted to preparatory and more or less standard definitions and results on intersection numbers of Cartier divisors, nef, big or ample Cartier divisors, and parameter spaces for morphisms from a fixed curve to a fixed variety, with various decorations.

• Chapters 3, 4, and 5 cover various aspects of the geometry of smooth projective varieties with many rational curves. At their heart lie Mori's bend-and-break lemmas in their different incarnations. We study varieties that are uniruled (i.e., covered by rational curves) or rationally connected (i.e., such that two general points can be joined by a rational curve). We give a structure theorem for varieties with nef anticanonical divisor, and prove that Fano varieties are rationally connected. The proof of this last result rests on a very general construction of Campana's, which in our case yields *the rational quotient*.

• In Chapters 6 and 7, we take the first steps toward Mori's minimal model program of classification of algebraic varieties by proving the cone and the contraction theorems. Given a variety X, the idea is to construct a nontrivial morphism $X \to Y$ with connected fibers to make X "simpler." Such a morphism is characterized by the set of numerical equivalence classes of the curves that it contracts, and the problem is to describe which subsets of classes in the cone $NE(X)$ generated by classes of curves on X can be obtained in this way. This is exactly what the contraction theorem does. The cone theorem gives a geometric description of (the closure of) the cone $NE(X)$. For smooth varieties, it is obtained as an application of Mori's bend-and-break techniques and undoubtedly represents the present culmination of the whole theory. Unfortunately, the contraction theorem is unattainable with these ideas. Also, it quickly becomes clear that we must allow some kind of singular varieties in the picture (the contraction of a smooth variety is often singular) and Mori's techniques break down in this case. We undertake a completely different approach (initiated by Kawamata) in the last chapter, which allows us to prove the cone and contraction theorems for varieties with canonical singularities (albeit in characteristic zero only). Alas, gone is the geometry, this is a world of cohomological calculations based on the whole machinery of vanishing theorems! This is by far the most difficult part of the book, but also a necessary evil justified by the scope of this new approach.

The "classical" way to state these vanishing theorems is complicated and makes them look rather unnatural. They look much better when they are rephrased within the so-called "logarithmic" framework, so I take this opportunity to briefly introduce the corresponding language. To include a complete treatment of singularities of pairs would, however, have taken me much too far and there already exist very thorough texts on this subtle question (such as [K7]), but I think the reader should know about the most simple-minded aspects of the theory. Similarly, most theorems concerning the minimal model program could also have been stated in the relative case. The proofs are not much more difficult, but I am not sure this degree of generality provides the best first contact with the theory, so I only mention without proofs some of the relevant results.

I have tried to keep the number of concepts that are used but not proved to a minimum (the construction of the space of morphisms and the Kawamata–Viehweg vanishing theorem are two examples). There is no doubt that some of the material presented here is not easy, but I have done my best to make it accessible. Except for some references to EGA in Chapter 5, my preferred source of reference has been Hartshorne's classic book [H1]. Like most of my predecessors, I claim—genuinely—that a reader familiar with [H1] will go through this book without too much trouble (but with work and patience). Behind the technicalities and the sometimes obscure terminology, I hope she (or he) will discover truly remarkable mathematics well worth her (or his) time.

Strasbourg, France   Olivier Debarre

Acknowledgments

This book is based on notes from a class taught at Harvard University where I was a visiting professor in the spring of 1999, and I would like to thank both the university for providing the support to make this possible and Joe Harris for the invitation. The comments of many of the people who attended this class were very helpful. I would like to thank in particular Brian Conrad, Andreas Gathman, Jacob Lurie, and Jason Starr. When I was working on my lecture notes to try to turn them into something readable, many people gave me advice or encouragement, including Marco Andreatta, Frédéric Campana, Jean-Louis Colliot-Thélène, Joe Harris, Yujiro Kawamata, David Madore, Francesco Polizzi, Michel Raynaud. Finally, János Kollár was always there to answer my questions; I also benefited immensely from the many books he wrote on the subject. To him go my special thanks. Thanks also to Raymond Séroul for his help with the illustrations, to Vincent Blanlœil for his explanations on computers, fonts, various species of PostScript and the like, and to the people at Springer-Verlag for their support. Most of all I want to thank my wife Becky for her unflagging enthusiasm for higher-dimensional algebraic geometry.

Strasbourg, France   Olivier Debarre

Contents

1
Curves and Divisors on Algebraic Varieties

In this preliminary chapter (as well as in the next), we gather standard material that will be used throughout the book. That is to say, material that dates mostly from pre-Mori times.

Most of the results obtained in this book are connected in some way or another with curves on algebraic varieties. Curves can be intersected with hypersurfaces and many of their properties depend only on the corresponding intersection numbers (we call them *numerical* properties). Consequently, we will say that two curves on an algebraic variety X are equivalent if they have the same intersection number with each hypersurface in X. It is standard to deal with this situation by injecting a bit of linear algebra into it: consider the real vector space with basis the family of all irreducible curves in X, and take the quotient by the subspace generated by the differences of any two equivalent curves. When X is proper, this quotient space $N_1(X)_{\mathbf{R}}$ *has finite dimension.*

There is more geometry attached to the situation: classes of irreducible curves in X generate a convex cone $\mathrm{NE}(X)$ in $N_1(X)_{\mathbf{R}}$, and it will be one of our aims to describe this cone (or rather its closure $\overline{\mathrm{NE}}(X)$) sitting in a finite-dimensional real vector space. This cone makes its first appearance in Kleiman's criterion, which says that a divisor is ample if and only if it has positive intersection with every nonzero element of $\overline{\mathrm{NE}}(X)$. We will prove this criterion, together with other basic results on ample, nef, and big divisors. A few examples will convince the reader that the geometry of this cone may be highly nontrivial.

Mori's approach to the classification of algebraic varieties makes heavy use of the easy remark, originally made by Hironaka, that many morphisms defined on a projective algebraic variety X are characterized by the curves they contract. Being contracted is a numerical property, and the classes of curves contracted by a given morphism generate a closed convex subcone of $\mathrm{NE}(X)$ which has the property of being *extremal*. This elementary remark raises in turn the important question of characterizing those extremal subcones of $\mathrm{NE}(X)$ that correspond to morphisms.

Why is this an important question? One way to start classifying projective algebraic varieties is to consider that if there is a morphism $\pi : X \to Y$, then Y is *simpler* than X. Starting from X and its cone of curves, one would therefore like to know which extremal subcones give rise to actual morphisms. Necessary and sufficient conditions are not yet known, nor are they anywhere in sight, but the contraction theorem will eventually give a partial but very satisfactory answer (for which we will have to wait until Chapter 6).

We briefly discuss this classification problem in the last section of this chapter and prove two simple results about the exceptional locus of a birational morphism $\pi : X \to Y$ (i.e., the closed subset of X where π is not a local isomorphism) when Y is smooth: it is covered by rational curves contracted by π and, when X also is smooth, there is a contracted rational curve on which the canonical class K_X has negative degree.

These two results show the importance of *rational* curves in the classification of algebraic varieties (they will occupy us for the greater part of Chapters 3 through 6).

We close the chapter with a few words about Mori's program: given a smooth projective variety X, one would like to construct, using contractions of extremal rays, a minimal model of X that behaves nicely, whose geometry is related to that of X, and which is as unique as possible. We are not in a position to prove anything new yet. The real work will only begin in the next chapter.

1.1. Conventions. All schemes are Noetherian[1] and separated. A proper scheme is a scheme of finite type and proper over a field. A variety is an integral scheme of finite type over a field. A subvariety is always closed, and so is a subscheme, unless otherwise mentioned.

1.1   Divisors and 1-Cycles

1.2. Cartier divisors. A Cartier divisor on a scheme X is a global section of the sheaf $\mathscr{K}_X^* / \mathscr{O}_X^*$, where $\mathscr{K}_X$ is the sheaf of total quotient rings of $\mathscr{O}_X$:

---

[1]Actually, this is not quite true: when we consider moduli spaces of morphisms from a curve to a variety (see Section 2.2), or Hilbert schemes (see 5.6), we will still call them "schemes" although they are only locally Noetherian (they have countably many Noetherian connected components). To stress the point, I will call them "locally Noetherian schemes" whenever possible. These will be the only conflicts with this convention.

on an open affine subset U of X, the ring $\mathscr{K}_X(U)$ is the localization of $\mathscr{O}_X(U)$ by the multiplicative system of elements that are not zero divisors and $\mathscr{K}_X^*(U)$ is the group of its invertible elements (if U is integral, $\mathscr{K}_X^*(U)$ is just the multiplicative group of the quotient field of $\mathscr{O}_X(U)$).

In other words, a Cartier divisor is given by a collection of pairs (U_i, f_i), where (U_i) is an open cover of X and f_i an invertible element of $\mathscr{K}_X(U_i)$, such that f_i/f_j is in $\mathscr{O}_X^*(U_i \cap U_j)$. When X is reduced, we may take integral open sets U_i, and f_i is then a nonzero rational function on U_i such that f_i/f_j is a regular function on $U_i \cap U_j$ that does not vanish.

A Cartier divisor is *principal* if it is in the image of the natural map $H^0(X, \mathscr{K}_X^*) \to H^0(X, \mathscr{K}_X^*/\mathscr{O}_X^*)$; in other words, when X is a variety, the divisor can be defined by a global nonzero rational function on the whole of X. Two Cartier divisors D and D' are *linearly equivalent* if their difference is principal; we write $D \equiv D'$. A Cartier divisor D is *effective* if it can be defined by a collection (U_i, f_i) where f_i is not a zero divisor in $\mathscr{O}_X(U_i)$; the f_i define a codimension 1 subscheme of X whose reduction is called the *support* of D. This sets up a one-to-one correspondence between effective Cartier divisors and subschemes locally defined by one equation which is not a zero divisor.

To a Cartier divisor D on X given by a collection (U_i, f_i), one can associate an invertible subsheaf $\mathscr{O}_X(D)$ of $\mathscr{K}_X$ by taking the sub-$\mathscr{O}_X$-module of $\mathscr{K}_X$ generated by $1/f_i$ on U_i. Every invertible subsheaf of $\mathscr{K}_X$ is obtained in this way, and two divisors are linearly equivalent if and only if their associated invertible sheaves are isomorphic ([H1], II, prop. 6.13). When X is integral, or projective over a field, every invertible sheaf is a subsheaf of $\mathscr{K}_X$ ([H1], II, rem. 6.14.1 and prop. 6.15).

Let $\pi : Y \to X$ be a morphism between schemes and let D be a Cartier divisor on X. The pull-back $\pi^*\mathscr{O}_X(D)$ is an invertible subsheaf of $\mathscr{K}_Y$, hence defines a divisor class on Y denoted by π^*D. Only the linear equivalence class of π^*D is well defined in general; however, when Y is reduced and D is an effective divisor (U_i, f_i) whose support contains the image of none of the irreducible components of Y, the collection $(\pi^{-1}(U_i), f_i \circ \pi)$ defines a divisor π^*D in that class. In particular, it makes sense to restrict a Cartier divisor to a subvariety not contained in its support, and to restrict a Cartier divisor *class* to any subvariety.

1.3. (Weil) divisors. Let X be a *normal* variety. For each integral hypersurface Y of X with generic point η, the local ring $\mathscr{O}_{\eta,X}$ has dimension 1 and is integrally closed, hence is a discrete valuation ring with valuation v_Y. For any nonzero rational function f on X, the integer $v_Y(f)$ is the order of vanishing of f along Y if it is nonnegative, and the opposite of the order of the pole of f along Y otherwise.

To a Cartier divisor D on X one can associate a finite linear combination

$$\sum_Y n_Y Y$$

of integral hypersurfaces of X: if D is given by a collection (U_i, f_i), the integer n_Y is the valuation of f_i along $Y \cap U_i$ for any i such that $Y \cap U_i$ is nonempty (it does not depend on the choice of such an i).

We define a *Weil divisor* (or simply a divisor) to be a finite formal linear combination of integral hypersurfaces with integral coefficients.

It will in fact be very useful to allow rational, and even real, coefficients; we will then talk about **Q**-divisors and **R**-divisors. On a normal variety, a Cartier divisor is therefore a locally principal divisor. We will say that a **Q**-divisor is **Q**-Cartier (sorry, but this is the current terminology) if some multiple is a Cartier divisor. A very important warning: it does not in general make sense to restrict a **Q**-divisor to a subvariety, even if it is **Q**-Cartier. We will do that only when the subvariety is not contained in the support of the divisor.

On a locally factorial variety (i.e., a variety whose local rings are unique factorization domains, e.g., a smooth variety), every divisor is locally principal ([H1], II, prop. 6.11), hence there is no distinction between Cartier divisors and Weil divisors. Up to Chapter 6 included, we will exclusively deal with smooth varieties, so these fine (but essential) details can be brushed aside for the moment.

1.4. Curves and 1-cycles. A curve is a 1-dimensional connected reduced *proper* scheme of finite type over a field **k**, not necessarily irreducible. Let X be a scheme of finite type over **k**. By a curve on X, we will mean either a *nonconstant* morphism from a (proper connected) curve with smooth irreducible components to X or its image on X. No confusion should arise from this ambiguity.

A 1-*cycle* on X is a formal sum $\sum_{i=1}^{s} n_i C_i$, where the n_i are integers and the C_i are integral curves on X; it is called *effective* if all the n_i are nonnegative.

1.2   Intersection Numbers

We fix a field **k**. All schemes considered in this section will be of finite type over **k**.

If X is a subscheme of $\mathbf{P}_\mathbf{k}^N$ of dimension n, it is proved in [H1], I, th. 7.5, that the function

$$m \mapsto \chi(X, \mathcal{O}_X(m))$$

is *polynomial of degree* n, i.e., takes the same values on the integers as a (uniquely determined) polynomial of degree n with rational coefficients, called the *Hilbert polynomial* of X. The degree of X in $\mathbf{P}^N$ is then defined as $n!$ times the coefficient of m^n. If H is a hyperplane section, the degree is also written as H^n. Our aim in this section is to generalize this definition and define an intersection number

$$D_1 \cdots D_n$$

for any Cartier divisors $D_1, \ldots, D_n$ on a proper n-dimensional scheme. We give a definition based on the same lines, using Euler characteristics. It has the advantage of being quick and efficient, but has very little geometric feeling to it.

Theorem 1.5 *Let $D_1, \ldots, D_r$ be Cartier divisors on a proper scheme X and let $\mathscr{F}$ be a coherent sheaf on X. The function*

$$(m_1, \ldots, m_r) \mapsto \chi(X, \mathscr{F}(m_1 D_1 + \cdots + m_r D_r))$$

takes the same values on $\mathbf{Z}^r$ as a polynomial with rational coefficients of degree at most the dimension of the support of $\mathscr{F}$.

PROOF. Any coherent sheaf $\mathscr{F}$ on a scheme X has a finite filtration

$$\mathscr{F} = \mathscr{F}_0 \supset \mathscr{F}_1 \supset \cdots \supset \mathscr{F}_m = 0$$

by coherent subsheaves such that each $\mathscr{F}_i/\mathscr{F}_{i+1}$ is torsion-free on an integral subscheme of X (see, e.g., [Gr2], III.3.1), so we may assume that X is integral and $\mathscr{F}$ is torsion-free.

By [Mat], th. 4.10, there exists a dense open subset U of X over which $\mathscr{F}$ is free of rank r. The choice of an isomorphism $\mathscr{F} \otimes \mathscr{O}_U \simeq \mathscr{O}_U^r$ gives an injection

$$\mathscr{F} \hookrightarrow \mathscr{K}_X^r$$

Set $\mathscr{G} = \mathscr{F} \cap \mathscr{O}_X^r$. In the following self-defining exact sequences

$$
\begin{array}{ccccccccc}
0 & \to & \mathscr{G} & \to & \mathscr{F} & \to & \mathscr{G}_1 & \to & 0 \\
0 & \to & \mathscr{G} & \to & \mathscr{O}_X^r & \to & \mathscr{G}_2 & \to & 0
\end{array}
\tag{1.1}
$$

the supports of the sheaves $\mathscr{G}_1$ and $\mathscr{G}_2$ do not meet U, hence have dimension less than X. By induction on the dimension of X, we are reduced to the case $\mathscr{F} = \mathscr{O}_X$ (and X integral).

Assume first $D_1 = \cdots = D_r = D$ and choose, as above, an embedding of $\mathscr{O}_X(D)$ into $\mathscr{K}_X$. Set

$$\mathscr{I}_1 = \mathscr{O}_X(-D) \cap \mathscr{O}_X \qquad\qquad \mathscr{I}_2 = \mathscr{O}_X(D) \cap \mathscr{O}_X$$

and let Y_j be the subscheme of X defined by the ideal $\mathscr{I}_j$. Since X is integral, Y_j has dimension smaller than X. Note that $\mathscr{I}_1(D)$ is isomorphic to $\mathscr{I}_2$, so there are exact sequences

$$
\begin{array}{ccccccccc}
0 & \to & \mathscr{I}_1(mD) & \to & \mathscr{O}_X(mD) & \to & \mathscr{O}_{Y_1}(mD) & \to & 0 \\
 & & \| & & & & & & \\
0 & \to & \mathscr{I}_2((m-1)D) & \to & \mathscr{O}_X((m-1)D) & \to & \mathscr{O}_{Y_2}((m-1)D) & \to & 0
\end{array}
\tag{1.2}
$$

which yield

$$\chi(X, mD) - \chi(X, (m-1)D) = \chi(Y_1, mD) - \chi(Y_2, (m-1)D)$$

By induction, the right-hand side of this equality is a polynomial function in m of degree $d < \dim(X)$. But if a function $f : \mathbf{Z} \to \mathbf{Z}$ is such that $m \mapsto f(m) - f(m-1)$ is polynomial of degree d, the function f itself is polynomial of degree $d + 1$ ([H1], I, prop. 7.3(b)); therefore, $\chi(X, mD)$ is a polynomial function in m of degree $\leq d + 1 \leq \dim(X)$, with rational coefficients.

We now treat the general case, still assuming $\mathscr{F} = \mathcal{O}_X$.

Lemma 1.6 *Let d be a positive integer and let $f : \mathbf{Z}^r \to \mathbf{Z}$ be a map such that for each $(n_1, \dots, n_{i-1}, n_{i+1}, \dots, n_r)$ in $\mathbf{Z}^{r-1}$, the map*

$$m \mapsto f(n_1, \dots, n_{i-1}, m, n_{i+1}, \dots, n_r)$$

is polynomial of degree at most d. The function f takes the same values as a polynomial in r indeterminates with rational coefficients.

PROOF. We proceed by induction on r, the case $r = 1$ being done in [H1], I, prop. 7.3(a). Assume $r > 1$; there exist functions $f_0, \dots, f_d : \mathbf{Z}^{r-1} \to \mathbf{Z}$ such that

$$f(m_1, \dots, m_r) = \sum_{j=0}^{d} f_j(m_1, \dots, m_{r-1})m_r^j$$

Pick distinct integers $c_0, \dots, c_d$. For each $i \in \{0, \dots, d\}$, there exists by the induction hypothesis a polynomial P_i with rational coefficients such that

$$f(m_1, \dots, m_{r-1}, c_i) = \sum_{j=0}^{d} f_j(m_1, \dots, m_{r-1})c_i^j = P_i(m_1, \dots, m_{r-1})$$

The matrix (c_i^j) is invertible and its inverse has rational coefficients. This proves that each f_j is a linear combination of $P_0, \dots, P_d$ with rational coefficients, hence the lemma.   □

From the case $r = 1$ and the lemma, we deduce that there exists a polynomial $P \in \mathbf{Q}[T_1, \dots, T_r]$ such that

$$\chi(X, m_1D_1 + \cdots + m_rD_r) = P(m_1, \dots, m_r)$$

for all integers $m_1, \dots, m_r$. Let d be its total degree, and let $n_1, \dots, n_r$ be integers such that the degree of the polynomial

$$Q(T) = P(n_1 T, \dots, n_r T)$$

is still d. Since

$$Q(m) = \chi(X, m(n_1 D_1 + \cdots + n_r D_r))$$

it follows from the case $r = 1$ that d is at most the dimension of X.   □

Definition 1.7 *Let $D_1, \ldots, D_r$ be Cartier divisors on a proper scheme X with $r \geq \dim(X)$. The intersection number*

$$D_1 \cdots D_r$$

is the coefficient of $m_1 \cdots m_r$ in the polynomial

$$\chi(X, m_1 D_1 + \cdots + m_r D_r)$$

Our next result (Proposition 1.8) shows that this number is an integer, and Theorem 1.5 that it vanishes for $r > \dim(X)$. If $D_1, \ldots, D_r$ are effective and meet properly in a finite number of points, the intersection number does have a geometric interpretation as the number of points in $D_1 \cap \cdots \cap D_r$, counted with multiplicity (see [K1], th. VI.2.8).

If Y is a subscheme of X of dimension at most s, we set

$$D_1 \cdots D_s \cdot Y = D_1|_Y \cdots D_s|_Y$$

If $\mathscr{F}$ is any coherent sheaf on X whose support has dimension at most r, we may also define an intersection number

$$D_1 \cdots D_r \cdot \mathscr{F}$$

as the coefficient of $m_1 \cdots m_r$ in the polynomial

$$\chi(X, \mathscr{F}(m_1 D_1 + \cdots + m_r D_r))$$

but this apparently greater generality is illusory.

Actually, once we have Kleiman's characterization of ample divisors in terms of their intersection numbers with 1-cycles, we will mostly intersect one divisor D with a curve C in X. In this case, things are very simple: the Riemann–Roch theorem ([H1], IV, th. 1.3)

$$\chi(C, mD) = m \deg(D) + \chi(C, \mathscr{O}_C)$$

implies

$$D \cdot C = \deg(\mathscr{O}_C(D)) \tag{1.3}$$

When D is a hypersurface that does not contain C, the intersection number therefore counts (with multiplicities) the number of points of intersection of D and C.

This formula implies in particular that the intersection number is linear in D (this holds in general by item (a) of the next proposition). We may extend the definition of the intersection number $D \cdot C$ by linearity for D a **Q**-Cartier **Q**-divisor and C a 1-cycle.

We now prove basic general properties of these intersection numbers.

Proposition 1.8 *Let $D_1, \ldots, D_n$ be Cartier divisors on a proper scheme X of dimension n.*

(a) *The map*

$$(D_1, \ldots, D_n) \mapsto D_1 \cdots D_n$$

is multilinear, symmetric, and takes integral values.

(b) *If D_n is effective with associated subscheme Y,*

$$D_1 \cdots D_n = D_1 \cdots D_{n-1} \cdot Y$$

PROOF. The map in (a) is symmetric by construction, but its multilinearity is not obvious. Assume $r \geq n$; the following identity

$$D_1 \cdots D_r = \sum_{I \subset \{1, \ldots, r\}} \varepsilon_I \, \chi\left(X, -\sum_{i \in I} D_i\right) \tag{1.4}$$

where $\varepsilon_I = (-1)^{\mathrm{Card}(I)}$, follows from the fact that if $P(T_1, \ldots, T_r)$ is any polynomial of total degree at most r, the coefficient of $T_1 \cdots T_r$ in P is

$$\sum_{I \subset \{1, \ldots, r\}} \varepsilon_I P(-m^I)$$

where $m_i^I = 1$ if $i \in I$ and 0 otherwise (this quantity vanishes for all other monomials of degree $\leq r$).

This identity shows that intersection numbers are integers. Also, the right-hand side of (1.4) vanishes for $r > n$, hence, for any divisors D_1, D_1', $D_2, \ldots, D_n$, the sum

$$\sum_{I \subset \{2, \ldots, n\}} \varepsilon_I \left(\chi\left(X, -\sum_{i \in I} D_i\right) - \chi\left(X, -D_1 - \sum_{i \in I} D_i\right) \right.$$
$$\left. - \chi\left(X, -D_1' - \sum_{i \in I} D_i\right) + \chi\left(X, -D_1 - D_1' - \sum_{i \in I} D_i\right) \right)$$

vanishes. On the other hand, $(D_1 + D_1') \cdot D_2 \cdots D_n$ is equal to

$$\sum_{I \subset \{2, \ldots, n\}} \varepsilon_I \left(\chi\left(X, -\sum_{i \in I} D_i\right) - \chi\left(X, -D_1 - D_1' - \sum_{i \in I} D_i\right) \right)$$

and $D_1 \cdot D_2 \cdots D_n + D_1' \cdot D_2 \cdots D_n$ to

$$\sum_{I \subset \{2, \ldots, n\}} \varepsilon_I \left(2\chi\left(X, -\sum_{i \in I} D_i\right) - \chi\left(X, -D_1 - \sum_{i \in I} D_i\right) - \chi\left(X, -D_1' - \sum_{i \in I} D_i\right) \right)$$

Putting all these identities together gives the desired equality

$$(D_1 + D_1') \cdot D_2 \cdots D_n = D_1 \cdot D_2 \cdots D_n + D_1' \cdot D_2 \cdots D_n$$

and proves (a).

In the situation of (b), we have

$$D_1 \cdots D_n = \sum_{I \subset \{1,\dots,n-1\}} \varepsilon_I \left(\chi\left(X, -\sum_{i \in I} D_i\right) - \chi\left(X, -D_n - \sum_{i \in I} D_i\right) \right)$$

From the exact sequence

$$0 \to \mathscr{O}_X\left(-D_n - \sum_{i \in I} D_i\right) \to \mathscr{O}_X\left(-\sum_{i \in I} D_i\right) \to \mathscr{O}_Y\left(-\sum_{i \in I} D_i\right) \to 0$$

we get

$$D_1 \cdots D_n = \sum_{I \subset \{1,\dots,n-1\}} \varepsilon_I \, \chi\left(Y, -\sum_{i \in I} D_i\right) = D_1 \cdots D_{n-1} \cdot Y$$

which proves (b). $\square$

1.9. Projection formula. Let $\pi : X \to Y$ be a proper morphism between varieties and let C be a curve on X. We define the 1-cycle $\pi_* C$ as follows: if C is contracted to a point by π, set $\pi_* C = 0$; if $\pi(C)$ is a curve on Y, set $\pi_* C = d\,\pi(C)$, where d is the degree of the morphism $C \to \pi(C)$ induced by π. If D is a Cartier divisor on X, we have the so-called *projection formula*

$$\pi^* D \cdot C = D \cdot \pi_* C \tag{1.5}$$

which is not hard to deduce from (1.3) and the fact that pulling back a divisor on a smooth curve by a finite morphism f multiplies its degree by $\deg(f)$ ([H1], II, prop. 6.9). This is a particular case of the following more general result. Recall that the degree of a dominant morphism $\pi : Y \to X$ between varieties is the degree of the field extension $\pi^* : K(X) \hookrightarrow K(Y)$ if this extension is finite, and 0 otherwise.

Proposition 1.10 (Projection formula) *Let $\pi : Y \to X$ be a surjective morphism between proper varieties. Let $D_1, \dots, D_r$ be Cartier divisors on X with $r \geq \dim(Y)$. We have*

$$\pi^* D_1 \cdots \pi^* D_r = \deg(\pi)(D_1 \cdots D_r)$$

PROOF. For any coherent sheaf $\mathscr{F}$ on Y, the sheaves $R^q \pi_* \mathscr{F}$ are coherent ([Gr2], th. 3.2.1) and there is a spectral sequence

$$H^p(X, R^q \pi_* \mathscr{F}) \Longrightarrow H^{p+q}(Y, \mathscr{F})$$

It follows that we have

$$\chi(Y, \mathscr{F}) = \sum_{q \geq 0} (-1)^q \chi(X, R^q \pi_* \mathscr{F})$$

Applying it to $\mathscr{F} = \mathscr{O}_Y(m_1 \pi^* D_1 + \cdots + m_r \pi^* D_r)$ and using the projection formula

$$R^q \pi_* \mathscr{F} \simeq R^q \pi_* \mathscr{O}_Y \otimes \mathscr{O}_Y(m_1 D_1 + \cdots + m_r D_r)$$

([Gr2], prop. 12.2.3), we get that $\pi^* D_1 \cdots \pi^* D_r$ is equal to the coefficient of $m_1 \cdots m_r$ in

$$\sum_{q \geq 0} (-1)^q \chi(X, R^q \pi_* \mathscr{O}_Y \otimes \mathscr{O}_X(m_1 D_1 + \cdots + m_r D_r))$$

If π is not generically finite, we have $r > \dim(X)$ and the coefficient of $m_1 \cdots m_r$ in each term of the sum vanishes by Theorem 1.5.

Otherwise, π is finite of degree d over a dense open subset U of Y, the sheaves $R^q \pi_* \mathscr{O}_Y$ have support outside of U for $q > 0$ ([H1], III, cor. 11.2) hence the coefficient of $m_1 \cdots m_r$ in the corresponding term vanishes for the same reason. Finally, $\pi_* \mathscr{O}_Y$ is free of degree d on some dense open subset of U and the formula follows using exact sequences as in (1.1). $\square$

1.3   The Cone of Curves

Let X be a proper scheme. We say that two Cartier divisors D and D' on X are *numerically equivalent* if they have same degree on every curve; we write $D \sim D'$. The quotient of the group of Cartier divisors by this equivalence relation is denoted by $N^1(X)_{\mathbf{Z}}$. We set

$$N^1(X)_{\mathbf{Q}} = N^1(X)_{\mathbf{Z}} \otimes \mathbf{Q} \qquad N^1(X)_{\mathbf{R}} = N^1(X)_{\mathbf{Z}} \otimes \mathbf{R}$$

These spaces are finite-dimensional vector spaces[2] and their dimension is called the *Picard number* of X, which we denote by ρ_X; we endow $N^1(X)_{\mathbf{R}}$ with its usual topology.

We say that a property of a divisor is *numerical* if it depends only on its numerical equivalence class.

1.11. Cone of curves. Two 1-cycles C and C' on X are *numerically equivalent* if they have the same intersection number with every Cartier divisor; we write $C \sim C'$. Call $N_1(X)_{\mathbf{Z}}$ the quotient group, and set

$$N_1(X)_{\mathbf{Q}} = N_1(X)_{\mathbf{Z}} \otimes \mathbf{Q} \qquad N_1(X)_{\mathbf{R}} = N_1(X)_{\mathbf{Z}} \otimes \mathbf{R}$$

---

[2]Over the complex numbers, this follows from Hodge theory : $N^1(X)_{\mathbf{Q}}$ is a subspace of $H^2(X, \mathbf{Q})$. The general argument is complicated ([Kle], p. 334).

The intersection pairing

$$N^1(X)_{\mathbf{Z}} \times N_1(X)_{\mathbf{Z}} \to \mathbf{Z}$$

is by definition nondegenerate. In particular, $N_1(X)_{\mathbf{R}}$ is a finite-dimensional real vector space which we endow with its usual topology. Inside $N_1(X)_{\mathbf{R}}$ sits the convex *cone of curves* $\mathrm{NE}(X)$, the set of classes of effective 1-cycles.

1.12. Relative cone of curves. Let X and Y be projective varieties. We define the relative cone of a morphism $\pi : X \to Y$ as the convex subcone $\mathrm{NE}(\pi)$ of $\mathrm{NE}(X)$ generated by the classes of curves contracted by π. Since Y is projective, an irreducible curve C on X is contracted by π if and only if $\pi_*[C] = 0$: being contracted is a numerical property. It follows that $\mathrm{NE}(\pi)$ is the intersection of $\mathrm{NE}(X)$ with the vector space $\mathrm{Ker}(\pi_*)$. It is therefore *closed* in $\mathrm{NE}(X)$ and *the class of an irreducible curve C lies in $\mathrm{NE}(\pi)$ if and only if this curve is contracted by π.*

What is this class of morphisms $\pi : X \to Y$ we have been talking about that are characterized by the curves they contract? A moment of thinking will convince the reader that this kind of information can only detect the connected components of the fibers, so we want to require at least connectedness of the fibers. When the characteristic of the base field is positive, this is not quite enough because of inseparability phenomena. The actual condition is

$$\pi_* \mathcal{O}_X \simeq \mathcal{O}_Y \tag{1.6}$$

1.13. If this condition holds, the fibers of π are indeed connected ([H1], III, cor. 11.3). For the converse, recall that any proper morphism $\pi : X \to Y$ has a *Stein factorization*

$$\pi : X \xrightarrow{\pi'} Y' \xrightarrow{g} Y$$

where Y' is the scheme $\mathbf{Spec}(\pi_*\mathcal{O}_Y)$ (for a definition, see [H1], II, ex. 5.17), so that $\pi'_*\mathcal{O}_X \simeq \mathcal{O}_{Y'}$, the morphism π' has connected fibers, and g is finite.

If the fibers of π are connected, the morphism g is bijective. *If the characteristic is zero,* g must be birational by generic smoothness ([H1], III, cor. 10.7). *Assume moreover that Y is normal;* if U is an affine open subset of Y, the ring $H^0(g^{-1}(U), \mathcal{O}_{Y'})$ is finite over the integrally closed ring $H^0(U, \mathcal{O}_Y)$, with the same quotient field, hence they are equal and g is an isomorphism. It follows that $\pi_*\mathcal{O}_X \simeq \mathcal{O}_Y$ by construction of the Stein factorization. In positive characteristic, g might very well be a bijection without being an isomorphism (even if Y is normal; think of the Frobenius morphism).

Finally, for any proper morphism $\pi : X \to Y$ with Stein factorization $\pi : X \xrightarrow{\pi'} Y' \to Y$, the curves contracted by π and the curves contracted by

π' are the same, hence the relative cones of π and π' are the same, so the condition (1.6) is really not too restrictive.

Our next result shows that morphisms π defined on a projective variety X that satisfy (1.6) are characterized by their relative cone $\mathrm{NE}(\pi)$. Moreover, this closed convex subcone of $\mathrm{NE}(X)$ has a simple geometric property: it is *extremal*, meaning that if a and b are in $\mathrm{NE}(X)$ and $a+b$ is in $\mathrm{NE}(\pi)$, both a and b are in $\mathrm{NE}(\pi)$ (geometrically, this means that $\mathrm{NE}(X)$ lies on one side of some hyperplane containing $\mathrm{NE}(\pi)$; we will prove this in Lemma 6.7, together with other elementary results on closed convex cones and their extremal subcones). As explained in the introduction to this chapter, it will be one of our aims to give sufficient conditions on an extremal subcone of $\mathrm{NE}(X)$ for it to be associated with an actual morphism, thereby converting geometric data on the (relatively) simple object $\mathrm{NE}(X)$ into information about the variety X.

Proposition 1.14 *Let X, Y, and Y' be projective varieties and let π : $X \to Y$ be a morphism.*

(a) *The subcone $\mathrm{NE}(\pi)$ of $\mathrm{NE}(X)$ is* extremal.

(b) *Assume $\pi_* \mathcal{O}_X \simeq \mathcal{O}_Y$ and let $\pi' : X \to Y'$ be another morphism.*

- *If $\mathrm{NE}(\pi)$ is contained in $\mathrm{NE}(\pi')$, there is a* unique *morphism $f : Y \to Y'$ such that $\pi' = f \circ \pi$.*

- *The morphism π is uniquely determined by $\mathrm{NE}(\pi)$ up to isomorphism.*

PROOF. Let $a = \sum a_i [C_i]$ and $a' = \sum a'_j [C'_j]$ be elements of $\mathrm{NE}(X)$, where a_i and a'_j are positive real numbers. If $a + a'$ is in $\mathrm{NE}(\pi)$, there exists a decomposition

$$\sum a_i [C_i] + \sum a'_j [C'_j] = \sum a''_k [C''_k]$$

where the C''_k are irreducible curves contracted by π and the a''_k are positive. Applying π_*, we get $\sum a_i \pi_*[C_i] + \sum a'_j \pi_*[C'_j] = 0$ in $N_1(Y)_{\mathbf{R}}$. Since Y is projective, the C_i and C'_j must be contracted by π hence a and a' are in $\mathrm{NE}(\pi)$. This proves (a).

To prove (b), we need the following rigidity result.

Lemma 1.15 *Let X, Y and Y' be varieties and let $\pi : X \to Y$ and π' : $X \to Y'$ be proper morphisms. Assume $\pi_* \mathcal{O}_X \simeq \mathcal{O}_Y$.*

(a) *If π' contracts one fiber $\pi^{-1}(y_0)$ of π, there is an open neighborhood Y_0 of y_0 in Y and a factorization*

$$\pi'|_{\pi^{-1}(Y_0)} : \pi^{-1}(Y_0) \xrightarrow{\pi} Y_0 \to Y'$$

(b) *If π' contracts each fiber of π, it factors through π.*

PROOF. Note that π is surjective. Let Z be the image of

$$g : X \xrightarrow{(\pi,\pi')} Y \times Y'$$

and let $p : Z \to Y$ and $p' : Z \to Y'$ be the two projections. Then $\pi^{-1}(y_0) = g^{-1}(p^{-1}(y_0))$ is contracted by π', hence by g. It follows that $p^{-1}(y_0) = g(g^{-1}(p^{-1}(y_0)))$ is a point, hence the proper surjective morphism p is finite over an open neighborhood Y_0 of y_0 in Y. Set $X_0 = f^{-1}(Y_0)$ and $Z_0 = p^{-1}(Y_0)$; we have $\mathcal{O}_{Z_0} \subset g_* \mathcal{O}_{X_0}$ and

$$\mathcal{O}_{Y_0} \subset p_* \mathcal{O}_{Z_0} \subset p_* g_* \mathcal{O}_{X_0} = \pi_* \mathcal{O}_{X_0} = \mathcal{O}_{Y_0}$$

hence $p_* \mathcal{O}_{Z_0} = \mathcal{O}_{Y_0}$. It follows that $p_0 : Z_0 \to Y_0$ is an isomorphism and $\pi' = p' \circ p_0^{-1} \circ \pi|_{X_0}$. This proves (a).

If π' contracts *each* fiber of π, the morphism p above is finite, one can take $Y_0 = Y$, and π' factors through π. This proves (b).   □

Going back to the proof of item (b) in the proposition, we assume now $\pi_* \mathcal{O}_X \simeq \mathcal{O}_Y$ and $\mathrm{NE}(\pi) \subset \mathrm{NE}(\pi')$. This means that every irreducible curve contracted by π is contracted by π', hence every (connected) fiber of π is contracted by π'. The existence of f follows from item (b) of the lemma. Assume $f' : Y \to Y'$ satisfies $\pi' = f' \circ \pi$. With the notation of the proof of the lemma, the composition $Z \xrightarrow{p} Y \xrightarrow{f'} Y'$ must be the second projection, hence $f' \circ p = p'$ and $f' = p' \circ p^{-1} = f$.

The second item in (b) follows from the first.   □

1.16. Mori's program. Part of this program aims at giving conditions under which a given extremal subcone V of $\mathrm{NE}(X)$ can be contracted. Once a contraction exists, its Stein factorization yields a contraction which is unique by the lemma and will be called *the* contraction map of V. We will have to wait until Theorem 7.39 to prove the existence of the contraction of some extremal subcones.

1.4   Ample and Very Ample Divisors

1.17. Ample Cartier divisors. A Cartier divisor D on a scheme X of finite type over a field is *ample* if, for every coherent sheaf $\mathcal{F}$ on X, there exists an integer m_0 such that $\mathcal{F}(mD)$ is generated by its global sections for all $m \geq m_0$. A Cartier divisor is ample if and only if some positive multiple is ample ([H1], II, prop. 7.5), hence it makes sense to talk about ample $\mathbf{Q}$-Cartier divisors, and even, on a normal variety, about $\mathbf{Q}$-Cartier $\mathbf{Q}$-divisors. To keep the text as readable as possible, whenever we say "ample $\mathbf{Q}$-divisor," or "ample divisor," it will always be understood that the divisor is **Q**-Cartier, and that the variety is normal if it is a **Q**-divisor.

1.18. Very ample Cartier divisors. A Cartier divisor D on a scheme X of finite type over a field is *very ample* if it is the restriction of a hyperplane for some immersion of X as a locally closed subset of $\mathbf{P}^N$. By a theorem of Serre, a very ample Cartier divisor is ample and conversely, some multiple of an ample divisor is very ample ([H1], II, th. 5.17 and 7.6). It follows that a proper scheme is projective if and only if it carries an ample divisor.

1.19. Another theorem of Serre says that a Cartier divisor D on a proper scheme X is ample if and only if, for every coherent sheaf $\mathscr{F}$ on X, there exists an integer m_0 such that $H^i(X, \mathscr{F}(mD)) = 0$ for all $m \geq m_0$ and all $i > 0$ ([H1], III, prop. 5.3). For X proper, it follows quite easily that the restriction of an ample Cartier divisor to a subscheme of X is ample, and that a Cartier divisor is ample on X if and only if it is ample on each irreducible component of X_{red}; also, the pull-back of an ample divisor by a finite morphism between proper schemes is ample ([H2], prop. I.4.2, I.4.3 and I.4.4).

1.20. Since the tensor product of two sheaves that are generated by their global sections has the same property, the sum of two ample Cartier divisors is ample and, if H is ample and if D is any Cartier divisor, $H + tD$ is ample for all t rational small enough. The same holds, of course, with **Q**-Cartier **Q**-divisors.

Similarly, using the Segre embedding, one sees that the sum of a very ample Cartier divisor and a Cartier divisor whose associated invertible sheaf is generated by its global sections is very ample. In particular, the sum of two very ample divisors is very ample and, if H is very ample and if D is any Cartier divisor, $mH + D$ is very ample for all integers m large enough.

1.5   The Nakai–Moishezon Ampleness Criterion

This is an ampleness criterion that involves only intersection numbers, but with all integral subschemes. Our aim is to prove eventually that ampleness is a *numerical property* in the sense that it depends only on intersection numbers with 1-cycles. This we will prove in Proposition 1.27(a).

Theorem 1.21 (Nakai–Moishezon criterion) *A Cartier divisor D on a proper scheme X is ample if and only if, for every integral subscheme Y of X, one has*

$$D^{\dim(Y)} \cdot Y > 0$$

The same result, of course, holds when D is a **Q**-Cartier **Q**-divisor. The criterion involves only *integral* subschemes because it is then easier to check, but if D is ample, $D^{\dim(Y)} \cdot Y$ is positive for *every* subscheme Y of X.

Having $D \cdot C > 0$ for every curve C on X does not in general imply that D is ample,[3] although there are some cases where it does (e.g., when $\mathrm{NE}(X)$ is closed, by Proposition 1.27(a); see 6.5 for a case where it happens).

PROOF OF THE THEOREM. One direction is easy: if D is ample, some positive multiple mD is very ample, hence defines an embedding $f : X \hookrightarrow \mathbf{P}^N$ such that $f^* \mathcal{O}_{\mathbf{P}^N}(1) \simeq \mathcal{O}_X(mD)$. In particular, for every integral subscheme Y of X,

$$(mD)^{\dim(Y)} \cdot Y = (mD|_Y)^{\dim(Y)} = \deg(f(Y)) > 0$$

The converse is more subtle. Let D be a Cartier divisor such that $D^{\dim(Y)} \cdot Y > 0$ for every integral subscheme Y of X. We show by induction on the dimension of X that D is ample on X. By 1.19, we may assume that X is integral.

As in the proof of Theorem 1.5, we choose an embedding of $\mathcal{O}_X(D)$ into $\mathcal{K}_X$, set $\mathcal{I}_1 = \mathcal{O}_X(-D) \cap \mathcal{O}_X$ and $\mathcal{I}_2 = \mathcal{O}_X(D) \cap \mathcal{O}_X$, and let Y_j be the subscheme[4] of X defined by the ideal $\mathcal{I}_j$; it has positive codimension in X. Consider the exact sequences (1.2)

$$0 \rightarrow \mathcal{I}_1(mD) \rightarrow \mathcal{O}_X(mD) \rightarrow \mathcal{O}_{Y_1}(mD) \rightarrow 0$$
$$\shortparallel$$
$$0 \rightarrow \mathcal{I}_2((m-1)D) \rightarrow \mathcal{O}_X((m-1)D) \rightarrow \mathcal{O}_{Y_2}((m-1)D) \rightarrow 0$$

By induction, D is ample on Y_1 and Y_2, hence $H^i(Y_j, mD)$ vanishes for $i > 0$ and all m sufficiently large. It follows that for $i \geq 2$,

$$h^i(X, mD) = h^i(X, \mathcal{I}_1(mD)) = h^i(X, \mathcal{I}_2((m-1)D)) = h^i(X, (m-1)D)$$

for all m sufficiently large. Since $D^{\dim(X)}$ is positive, $\chi(X, mD)$ goes to infinity with m. It follows that

$$h^0(X, mD) - h^1(X, mD)$$

hence also $h^0(X, mD)$, go to infinity with m. To prove that D is ample, we may replace it with any positive multiple. So we may assume that D is effective. The exact sequence

$$0 \rightarrow \mathcal{O}_X((m-1)D) \rightarrow \mathcal{O}_X(mD) \rightarrow \mathcal{O}_D(mD) \rightarrow 0$$

---

[3]See 1.35 for a counterexample with $D^2 = 0$ on a ruled surface, and [H2], p. 57, for a counterexample with $D^3 > 0$ on a projective threefold.

[4]The reader only interested in the case where X is projective will use 1.18 to prove that there exist effective divisors D_1 and D_2 such that D is linearly equivalent to $D_1 - D_2$, and use instead of Y_j the codimension 1 subscheme of X associated with the divisor D_j. In that case, the ideal sheaf $\mathcal{I}_j$ is the invertible sheaf $\mathcal{O}_X(-D_j)$.

and the vanishing of $H^1(D, mD)$ for all m sufficiently large yield a surjection

$$\rho_m : H^1(X, (m-1)D) \to H^1(X, mD)$$

The dimensions $h^1(X, mD)$ form a nonincreasing sequence of numbers which must eventually become stationary, in which case ρ_m is bijective and the restriction

$$H^0(X, mD) \to H^0(D, mD)$$

is surjective. By induction, D is ample on D, hence $\mathcal{O}_D(mD)$ is generated by its global sections for all m sufficiently large. It follows that $|mD|$ has no base-point on D. Since it has a section that vanishes exactly on the support of D, it has no base-point at all, hence defines a proper morphism f from X to a projective space $\mathbf{P}^N$. Since D has positive degree on every curve, f has finite fibers, hence is finite. Since $\mathcal{O}_X(D) = f^* \mathcal{O}_{\mathbf{P}^N}(1)$, the conclusion follows from 1.19.   □

1.6   Nef Divisors

In view of the Nakai–Moishezon criterion, it is natural to make the following definition: a Cartier divisor D on a proper scheme X is *nef*[5] if it satisfies, for every integral subscheme Y of X of dimension r,

$$D^r \cdot Y \geq 0 \qquad\qquad (1.7)$$

The beginning of the proof of Theorem 1.5 shows that (1.7) holds for every integral subscheme Y of X of dimension r if and only if the inequality

$$D^r \cdot \mathcal{F} \geq 0$$

holds for every coherent sheaf $\mathcal{F}$ on X whose support has dimension at most r, hence to (1.7) for *every* subscheme Y of X of dimension r. Also, it implies that D is nef if and only if its restriction to each irreducible component of X_{red} is. The restriction of a nef divisor to a subscheme is again nef.

This definition still makes sense for $\mathbf{Q}$-Cartier divisors, and even, on a normal variety, for $\mathbf{Q}$-Cartier $\mathbf{Q}$-divisors (see also 1.28). As for ample divisors, whenever we say "nef $\mathbf{Q}$-divisor," or "nef divisor," it will always be understood that the divisor is $\mathbf{Q}$-Cartier and that the variety is normal if it is a $\mathbf{Q}$-divisor.

1.22. Sum of ample and nef divisors. Let us begin with a lemma that will be used repeatedly in what follows.

---

[5]This acronym comes from "numerically effective." This definition has been adapted by Demailly, Peternell, and Schneider in [DPS] to the case of Kähler manifolds, which might have no curves at all, by requiring certain positivity properties of the curvature.

Lemma 1.23 *Let X be a proper scheme of dimension n, let D be a Cartier divisor, and let H be an ample divisor on X. If $D^r \cdot Y \geq 0$ for every integral subscheme Y of X of dimension r, we have*

$$D^r \cdot H^{n-r} \geq 0$$

PROOF. We proceed by induction on n. As above, the inequality $D^r \cdot H^{n-r} \geq 0$ holds on X if and only if it does on each irreducible component of X_{red}. We may therefore assume that X is integral and $r < n$.

Let m be an integer such that mH is very ample. The linear system $|mH|$ contains an effective divisor Y. Using Proposition 1.8(b), we get

$$
\begin{aligned}
D^r \cdot H^{n-r} &= \frac{1}{m} D^r \cdot H^{n-r-1} \cdot (mH) \\
&= \frac{1}{m} D^r \cdot H^{n-r-1} \cdot Y \\
&= \frac{1}{m} (D|_Y)^r \cdot (H|_Y)^{n-r-1}
\end{aligned}
$$

and this is nonnegative by the induction hypothesis.   □

Let now X be a projective scheme, let D be a nef divisor on X, let H be an ample divisor, and let Y be an r-dimensional subscheme of X. Since $D|_Y$ is nef, the lemma implies

$$D^s \cdot H^{r-s} \cdot Y \geq 0 \qquad (1.8)$$

for $0 \leq s \leq r$, hence

$$(D+H)^r \cdot Y = H^r \cdot Y + \sum_{s=1}^{r} \binom{r}{s} D^s \cdot H^{r-s} \cdot Y \geq (H|_Y)^r > 0$$

because $H|_Y$ is ample. By the Nakai–Moishezon criterion, $D + H$ is ample: *on a projective scheme, the sum of a nef divisor and an ample divisor is ample.* This still holds for **Q**-Cartier **Q**-divisors.

1.24. Sum of nef divisors. Let D and E be nef divisors on a projective scheme X of dimension n, and let H be an ample divisor on X. We just saw that $E + tH$ is ample for all positive rationals t and that, for every subscheme Y of X of dimension r, we have

$$D^{r-s} \cdot (E+tH)^s \cdot Y > 0$$

for $0 \leq s \leq r$. By letting t go to 0, we get, using multilinearity,

$$D^{r-s} \cdot E^s \cdot Y \geq 0$$

and by the binomial formula (using Proposition 1.8(a)),

$$(D+E)^r \cdot Y \geq 0$$

It follows that $D + E$ is nef: on a projective scheme, a sum of nef divisors is nef. This is still true on a proper scheme by Theorem 1.26.

Moreover, we have $D^{n-s} \cdot E^s \geq 0$ for $0 \leq s \leq n$, hence $(D + E)^n \geq D^n$ by the binomial formula again. This of course remains true for $\mathbf{Q}$-Cartier $\mathbf{Q}$-divisors.

1.25. A numerical characterization of nefness. Nefness is much more well-behaved than ampleness; for example, the following fundamental result says that it is enough to check nonnegativity on curves. We will use it constantly. It implies by the projection formula (1.5) that the inverse image of a nef divisor by any morphism is still nef.

Theorem 1.26 *Let X be a proper scheme. A Cartier divisor on X is nef if and only if it has nonnegative intersection with every curve on X.*

The same result of course holds for $\mathbf{Q}$-Cartier $\mathbf{Q}$-divisors.

PROOF OF THE THEOREM. We may assume that X is integral. Let D be a Cartier divisor on X with nonnegative degree on every curve, and let $n = \dim(X)$. Proceeding by induction on n, it is enough to prove $D^n \geq 0$. By Chow's lemma, there exist a projective variety X' and a birational surjective morphism $\pi : X' \to X$. By the projection formula (1.5), the divisor $\pi^* D$ has nonnegative intersection with every curve on X'. Since $D^n = (\pi^* D)^n$ (Proposition 1.10), we may (and will) assume that X is projective.

Let H be an ample divisor on X and set $D_t = D + tH$. Consider the degree n polynomial

$$P(t) = D_t^n = D^n + \binom{n}{1}(D^{n-1} \cdot H)t + \cdots + H^n t^n$$

We need to show $P(0) \geq 0$. Assume the contrary. Since the leading coefficient of P is positive, it has a largest positive real root t_0 and $P(t) > 0$ for $t > t_0$.

For every subvariety Y of X of positive dimension $r < n$, the divisor $D|_Y$ is nef by induction. By (1.8), we have

$$D^s \cdot H^{r-s} \cdot Y \geq 0$$

for $0 \leq s \leq r$. Also, $H^r \cdot Y > 0$ because H is ample. This implies, for $t > 0$,

$$D_t^r \cdot Y = D^r \cdot Y + \binom{r}{1}(D^{r-1} \cdot H \cdot Y)t + \cdots + (H^r \cdot Y)t^r > 0 \qquad (1.9)$$

Since $D_t^n = P(t) > 0$ for $t > t_0$, the Nakai–Moishezon criterion implies that D_t is ample for t rational and $t > t_0$.

Note that P is the sum of the polynomials

$$Q(t) = D_t^{n-1} \cdot D \qquad \text{and} \qquad R(t) = tD_t^{n-1} \cdot H$$

Since D_t is ample for t rational $> t_0$ and D has nonnegative degree on curves, we have $Q(t) \geq 0$ for *all* $t \geq t_0$ by Lemma 1.23. By the same lemma, (1.9) implies

$$D^r \cdot H^{n-r} \geq 0$$

for $0 \leq r < n$, hence

$$R(t_0) = (D^{n-1} \cdot H)t_0 + \binom{n-1}{1}(D^{n-2} \cdot H^2)t_0^2 + \cdots + H^n t_0^n \geq H^n t_0^n > 0$$

We get the contradiction

$$0 = P(t_0) = Q(t_0) + R(t_0) \geq R(t_0) > 0$$

This proves that $P(t)$ does not vanish for $t > 0$, hence

$$0 \leq P(0) = D^n$$

This proves the theorem.   □

1.7   A Numerical Characterization of Ampleness

We have now gathered enough material to prove our main characterization of ample divisors. It has numerous implications, the most obvious being that ampleness is a numerical property, so we can talk about ample classes in $N^1(X)_{\mathbf{Q}}$. These classes form an open cone (by 1.19) in $N^1(X)_{\mathbf{Q}}$, called the ample cone, whose closure is the nef cone (by 1.22 and Theorem 1.26).

The criterion also implies that the closed cone of curves of a *projective* variety contains no lines: by Lemma 6.7(a), a closed convex cone contains no lines if and only if it is contained in an open half-space plus the origin. The result does not hold for arbitrary proper X, but it does hold for proper smooth schemes ([K1], th. VI.2.19). Therefore, a smooth proper variety is projective if and only if its closed cone of curves contains no lines.

Theorem 1.27 (Kleiman's criterion) *Let X be a projective variety.*

(a) *A Cartier divisor D on X is ample if and only if $D \cdot z > 0$ for all nonzero z in $\overline{\mathrm{NE}}(X)$.*

(b) *For any ample divisor H and any integer k, the set $\{z \in \overline{\mathrm{NE}}(X) \mid H \cdot z \leq k\}$ is compact, hence contains only finitely many classes of irreducible curves.*

Item (a) of course still holds when D is a $\mathbf{Q}$-Cartier $\mathbf{Q}$-divisor.

PROOF OF THE THEOREM. Assume D is ample. Clearly, one has $D \cdot z \geq 0$ for all z in $\overline{\mathrm{NE}}(X)$. Assume $D \cdot z = 0$ and $z \neq 0$. Since the intersection pairing is nondegenerate, there exists a divisor E such that $E \cdot z < 0$, hence $(D + tE) \cdot z < 0$ for all positive t. In particular, $D + tE$ cannot be ample, which contradicts 1.20.

Assume for the converse that D is positive on $\overline{\mathrm{NE}}(X) - \{0\}$. Choose a norm $\| \cdot \|$ on $N_1(X)_{\mathbf{R}}$. The set

$$K = \{ z \in \overline{\mathrm{NE}}(X) \mid \|z\| = 1 \}$$

is compact. The functional $z \mapsto D \cdot z$ is positive on K, hence is bounded from below by a positive rational number a. Let H be an ample divisor on X. The functional $z \mapsto H \cdot z$ is bounded from above on K by a positive rational number b. It follows that $D - \frac{a}{b}H$ is nonnegative on K, hence on the cone $\overline{\mathrm{NE}}(X)$. This is exactly saying that $D - \frac{a}{b}H$ is nef, and by 1.22,

$$D = \left(D - \frac{a}{b}H \right) + \frac{a}{b}H$$

is ample. This proves (a).

Let $D_1, \ldots, D_r$ be Cartier divisors on X such that $([D_1], \ldots, [D_r])$ is a basis for $N_1(X)_{\mathbf{Q}}$. There exists an integer m such that $mH \pm D_i$ is ample for each i in $\{1, \ldots, r\}$. For any z in $\overline{\mathrm{NE}}(X)$, we then have $(mH \pm D_i) \cdot z \geq 0$ hence $|D_i \cdot z| \leq mH \cdot z$. If $H \cdot z \leq k$, this bounds the coordinates of z and defines a closed bounded set. It contains at most finitely many classes of irreducible curves, because the set of this classes is by construction discrete in $N_1(X)_{\mathbf{R}}$. □

1.28. Nef R-divisors. On a normal variety, the definition (1.7) of nef divisors also makes sense for linear combinations with real coefficients of Cartier divisors. We will occasionally use these divisors, which we will call *nef R-divisors*. Let D be a linear combination with real coefficients of Cartier divisors on a projective normal variety X. Assume that D is non-negative on curves and let H be an ample divisor on X. Choose a norm $\| \cdot \|$ on $N_1(X)_{\mathbf{R}}$. The set

$$K = \{ z \in \overline{\mathrm{NE}}(X) \mid \|z\| = 1 \}$$

is compact. By Kleiman's criterion, for all $t > 0$, the **R**-divisor $D + tH$ is positive on K, hence can be approximated by a **Q**-Cartier **Q**-divisor with the same property. It follows that $[D]$ is a limit of classes of **Q**-divisors which are positive on K, hence ample.

This implies that properties in 1.24, as well as Theorem 1.26, are still valid for divisors which are linear combinations with real coefficients of Cartier divisors on a projective normal variety X.

1.29. Nef and big divisors. A *nef* Cartier divisor D on a proper scheme of dimension n is *big* if D^n is positive. This definition still makes sense for a $\mathbf{Q}$-Cartier $\mathbf{Q}$-divisor. By 1.24, the sum of a nef and big $\mathbf{Q}$-divisor and of a nef $\mathbf{Q}$-divisor is nef and big.

Ample divisors are nef and big, but not conversely (see footnote 3, p. 15). Nef and big divisors share many of the properties of ample divisors: for example, the next proposition shows that the dimension of the space of sections of their successive multiples grows in the same fashion. We will see much later in Chapter 7 that many vanishing theorems, classically known to hold for ample divisors, are still valid for nef and big divisors. They are, however, much more tractable. For instance, the pull-back of a nef and big divisor by a generically finite morphism is still nef and big.

1.8   An Asymptotic Form of Riemann–Roch

We know from Theorem 1.5 that the growth of the Euler characteristic $\chi(X, mD)$ of successive multiples of a divisor D on a proper scheme X of dimension n is polynomial in m with leading coefficient $D^n/n!$. The full Riemann–Roch theorem identifies the coefficients of that polynomial.

We study here the dimensions $h^0(X, mD)$ and show that they grow in general not faster than some multiple of m^n and exactly like $\chi(X, mD)$ when D is nef (this is obvious when D is ample because $h^i(X, mD)$ vanishes for $i > 0$ and all m big enough by 1.18). Item (b) is particularly useful when D is in addition big.

1.30.   The corresponding estimation

$$\liminf_{m \to +\infty} \frac{h^0(X, mD)}{m^n} > 0$$

is actually the definition of a big divisor, without the requirement that it be nef.

Proposition 1.31 *Let D be a Cartier divisor on a proper scheme X of dimension n and let $\mathscr{F}$ be a coherent sheaf on X.*

(a) *We have*
$$h^i(X, \mathscr{F}(mD)) = O(m^n)$$
   for all i.

(b) *If D is nef, we have*
$$h^i(X, \mathscr{F}(mD)) = O(m^{n-1})$$
   for all positive i, hence
$$h^0(X, mD) = m^n \frac{D^n}{n!} + O(m^{n-1})$$

PROOF. To prove (a) and (b) we may, as in the proof of Theorem 1.5, assume X integral and $\mathscr{F} = \mathscr{O}_X$. Keeping the same notation

$$\mathscr{I}_1 = \mathscr{O}_X(-D) \cap \mathscr{O}_X \qquad\qquad \mathscr{I}_2 = \mathscr{O}_X(D) \cap \mathscr{O}_X$$

we have the same exact sequences (1.2)

$$
\begin{array}{ccccccc}
0 \to & \mathscr{I}_1(mD) & \to & \mathscr{O}_X(mD) & \to & \mathscr{O}_{Y_1}(mD) & \to 0 \\
& \| & & & & & \\
0 \to & \mathscr{I}_2((m-1)D) & \to & \mathscr{O}_X((m-1)D) & \to & \mathscr{O}_{Y_2}((m-1)D) & \to 0
\end{array}
$$

where Y_j is the subscheme of X defined by the ideal $\mathscr{I}_j$. The long exact sequences in cohomology give

$$
\begin{aligned}
h^i(X, mD) &\leq h^i(X, \mathscr{I}_1(mD)) + h^i(Y_1, mD) \\
&= h^i(X, \mathscr{I}_2((m-1)D)) + h^i(Y_1, mD) \\
&\leq h^i(X, (m-1)D) + h^{i-1}(Y_2, (m-1)D) + h^i(Y_1, mD)
\end{aligned}
$$

To prove (a) and (b), we proceed by induction on n. These inequalities imply, with the induction hypothesis,

$$h^i(X, mD) \leq h^i(X, (m-1)D) + O(m^{n-1})$$

and (a) follows by summing up these inequalities. If D is nef, we get in the same way, for $i \geq 2$,

$$h^i(X, mD) \leq h^i(X, (m-1)D) + O(m^{n-2})$$

hence $h^i(X, mD) = O(m^{n-1})$. This implies in turn, by the very definition of D^n,

$$
\begin{aligned}
h^0(X, mD) - h^1(X, mD) &= \chi(X, mD) + O(m^{n-1}) \\
&= m^n \frac{D^n}{n!} + O(m^{n-1})
\end{aligned}
$$

If $h^0(X, mD) = 0$ for all $m > 0$, the left-hand side of this equality is nonnegative. Since D^n is nonnegative, it must be 0 and $h^1(X, mD) = O(m^{n-1})$.

Otherwise, there exists an effective divisor Y in some linear system $|m_0 D|$ and the exact sequence

$$0 \to \mathscr{O}_X((m-m_0)D) \to \mathscr{O}_X(mD) \to \mathscr{O}_Y(mD) \to 0$$

yields

$$
\begin{aligned}
h^1(X, mD) &\leq h^1(X, (m-m_0)D) + h^1(Y, mD) \\
&= h^1(X, (m-m_0)D) + O(m^{n-2})
\end{aligned}
$$

by induction. Again, $h^1(X, mD) = O(m^{n-1})$ and (b) is proved.   □

Corollary 1.32 *Let D be a nef and big $\mathbf{Q}$-divisor on a projective variety X. There exists an effective $\mathbf{Q}$-Cartier $\mathbf{Q}$-divisor E on X such that $D - tE$ is ample for all rationals t in $(0, 1]$.*

PROOF. We may assume that D has integral coefficients. Let n be the dimension of X and let H be an effective ample divisor on X. Since $h^0(H, mD) = O(m^{n-1})$, we have $H^0(X, mD - H) \neq 0$ for all m sufficiently large by the proposition. Writing $mD \equiv H + E'$, with E' effective, we get

$$D = \left(\frac{t}{m}H + (1-t)D\right) + \frac{t}{m}E'$$

where $\frac{t}{m}H + (1-t)D$ is ample for all rationals t in $(0, 1]$ by 1.22. This proves the corollary with $E = \frac{1}{m}E'$. $\qquad\square$

1.9 Examples

We will explore on some examples the structure of the cone of curves of some projective varieties and the correspondence between morphisms and extremal subcones.

1.33. One-dimensional cone of curves. On a projective variety X for which the vector space $N_1(X)_{\mathbf{R}}$ has dimension 1 (such as a projective space), there is not much to say: there are only two extremal subcones in $\mathrm{NE}(X)$, to wit $\mathrm{NE}(X)$ itself and $\{0\}$. They correspond, respectively, to the morphism from X to a point, and to the identity morphism $X \to X$.

1.34. Product of projective spaces. Let X be a product $\mathbf{P} \times \mathbf{P}'$ of two projective spaces. The vector space $N_1(X)_{\mathbf{R}}$ has dimension 2. Let ℓ be the class of a line in $\mathbf{P}$ and let ℓ' be the class of a line in $\mathbf{P}'$. The cone of curves of X is

$$\mathrm{NE}(X) = \mathbf{R}^+\ell + \mathbf{R}^+\ell'$$

There are only two proper extremal subcones, the rays generated by ℓ and ℓ'. They correspond, respectively, to the projections $X \to \mathbf{P}'$ and $X \to \mathbf{P}$.

1.35. Ruled surfaces. Let X be a $\mathbf{P}^1$-bundle over a smooth curve C of genus g. We will freely use facts about ruled surfaces proved in [H1], V, §2 as well as the notation thereof. The vector space $N^1(X)_{\mathbf{R}}$ has dimension 2. It is generated by the class of a fiber F and the class of a certain section C_0. The ray generated by the class $[F]$ is extremal since it is the relative subcone associated with the projection $\pi : X \to C$.

Set $e = -C_0^2$. This is an invariant of X which can take any value $\geq -g$.

When $e \geq 0$, any irreducible curve on X is numerically equivalent to C_0 or to $aC_0 + bF$, with $a \geq 0$ and $b \geq ae$. In particular,

$$\mathrm{NE}(X) = \overline{\mathrm{NE}}(X) = \mathbf{R}^+[C_0] + \mathbf{R}^+[F]$$

We have already seen that the extremal ray $\mathbf{R}^+[F]$ is associated with the projection $X \to C$. The extremal ray $\mathbf{R}^+[C_0]$ may or may not be contracted. When $g = 0$, the contraction exists: it is the morphism associated with the base-point-free linear system $|C_0 + eF|$, and, when $e > 0$, it is birational and contracts only the curve C_0 to a point on a surface (when $e = 0$, the surface X is $\mathbf{P}^1 \times \mathbf{P}^1$).

When $e < 0$ (so in particular $g > 0$) and the characteristic is zero, any irreducible curve on X is numerically equivalent to C_0 or to $aC_0 + bF$, with $a \geq 0$ and $2b \geq ae$. Furthermore, any divisor $aC_0 + bF$, with $a > 0$ and $2b > ae$, is ample, hence some multiple is the class of a curve. This implies

$$\overline{\mathrm{NE}}(X) = \mathbf{R}^+[2C_0 + eF] + \mathbf{R}^+[F]$$

When $g = 1$, the only possible value is $e = -1$ and there is a curve numerically equivalent to $2C_0 + eF$: the cone $\mathrm{NE}(X)$ is closed.

However, when $g \geq 2$ and the base field is $\mathbf{C}$, there exists a rank-2 vector bundle $\mathscr{E}$ of degree 0 on C all of whose symmetric powers are *stable*.[6] The normalization of $\mathscr{E}$ in the sense of [H1], V, prop. 2.8, has even positive degree $-e$. For the associated ruled surface $X = \mathbf{P}(\mathscr{E})$, no multiple of the class $[2C_0 + eF]$ is effective.[7] In particular, it is not ample, although it has positive intersection with every curve on X. By Kleiman's criterion, this implies that the cone $\mathrm{NE}(X)$ is *not* closed. Actually,

$$\mathrm{NE}(X) = \left(\mathbf{R}^+[2C_0 + eF] + \mathbf{R}^{+*}[F]\right) \cup \{0\}$$

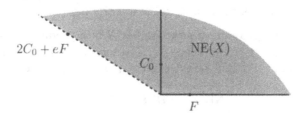

The nonclosed effective cone of the ruled surface X.

1.36. A useful example. The following constructions will provide useful examples in a variety of contexts.

Let r and s be *positive* integers, let $\mathscr{E}$ be the vector bundle on $\mathbf{P}^s$ associated with the locally free sheaf $\mathscr{O}_{\mathbf{P}^s} \oplus \mathscr{O}_{\mathbf{P}^s}(1)^{r+1}$, and let $Y_{r,s}$ be the smooth

---

[6] For the definition of stability and the construction of $\mathscr{E}$, see [H2], I, §10.

[7] The class $[C_0 + (e/2)F]$ is that of the line bundle $\mathscr{O}_X(1)$ on X associated with the bundle $\mathscr{E}$. For any divisor D on C of degree e, the vector space $H^0(X, \mathscr{O}_X(a(2C_0 + \pi^*D)))$ is therefore isomorphic to $H^0(C, S^a\mathscr{E}(D'))$, where D' has degree 0, and this space vanishes because $S^a\mathscr{E}(D')$ is stable of degree 0.

$(r + s + 1)$-dimensional variety $\mathbf{P}(\mathscr{E})$.[8] The projection $\pi_{r,s} : Y_{r,s} \to \mathbf{P}^s$ has a section σ with image $P_{r,s}$ (or simply P) corresponding to the trivial quotient of $\mathscr{E}$. The vector space $N^1(Y_{r,s})_{\mathbf{R}}$ has dimension 2 and is generated by the class ξ of the line bundle $\mathscr{O}_{Y_{r,s}}(1)$ and the inverse image h of the hyperplane class in $\mathbf{P}^s$. The restriction of ξ to P is zero ([H1], V, prop. 2.3 and 2.6).

Let ℓ be the class in $Y_{r,s}$ of a line in P, and let ℓ' be the class of a line in a fiber of $\pi_{r,s}$. We have

$$h \cdot \ell = 1 , \quad h \cdot \ell' = 0 , \quad \xi \cdot \ell = 0 , \quad \xi \cdot \ell' = 1$$

Since ξ and h are *nef*, it implies (for $s > 0$)

$$\mathrm{NE}(Y_{r,s}) = \overline{\mathrm{NE}}(Y_{r,s}) = \mathbf{R}^+\ell + \mathbf{R}^+\ell'$$

Let now $\varepsilon_1 : X_{r \cdot s} \to Y_{r,s}$ be the blow-up of P. Since the normal bundle to P in $Y_{r,s}$ is $\mathscr{O}_P(-1)^{r+1}$, the exceptional divisor E is isomorphic to $\mathbf{P}^r \times \mathbf{P}^s$ and the line bundle $\mathscr{O}_E(E)$ is of type $(-1, -1)$.[9] The vector space $N^1(X_{r \cdot s})_{\mathbf{R}}$ has dimension 3, generated by the (pull-backs of the) nef classes ξ and h, and by $[E]$. Let ℓ be the class of a line in E that projects onto a line in P, let ℓ'' be the class of a line in E contracted by the blow-up, and let ℓ' be the class of the strict transform of a line in a fiber of $\pi_{r,s} : Y_{r,s} \to \mathbf{P}^s$ that meets P. We have the following multiplication table

$$
\begin{array}{ccc}
h \cdot \ell = 1 & h \cdot \ell' = 0 & h \cdot \ell'' = 0 \\
\xi \cdot \ell = 0 & \xi \cdot \ell' = 1 & \xi \cdot \ell'' = 0 \\
[E] \cdot \ell = -1 & [E] \cdot \ell' = 1 & [E] \cdot \ell'' = -1
\end{array}
$$

Let $a\ell + a'\ell' + a''\ell''$ be the class of an irreducible curve C contained in $X_{r \cdot s}$. The subvariety P of $Y_{r,s}$ is defined as the complete intersection of $r + 1$ divisors with class $\xi - h$. This implies that if C is not contained in E, the class $\xi - h - [E]$ on $X_{r \cdot s}$ has nonnegative intersection with C, hence

$$a = H \cdot C \geq 0 , \quad a' = \xi \cdot C \geq 0 , \quad a'' = (\xi - h - [E]) \cdot C \geq 0$$

If C is contained in E, we have $a' = 0$, and a and a'' are nonnegative. All this implies

$$\mathrm{NE}(X_{r \cdot s}) = \overline{\mathrm{NE}}(X_{r \cdot s}) = \mathbf{R}^+\ell + \mathbf{R}^+\ell' + \mathbf{R}^+\ell'' \qquad (1.10)$$

The variety $X_{r \cdot s}$ can also be described as $\mathbf{P}(\mathscr{O}_{\mathbf{P}^r \times \mathbf{P}^s} \oplus \mathscr{O}_{\mathbf{P}^r \times \mathbf{P}^s}(1, 1))$. It is in particular isomorphic to $X_{s \cdot r}$. There is a commutative diagram

$$
\begin{array}{ccccc}
Y_{s,r} & \xleftarrow{\varepsilon_2} & X_{r \cdot s} & \xrightarrow{\varepsilon_1} & Y_{r,s} \\
\downarrow{\scriptstyle \pi_{s,r}} & & \downarrow{\scriptstyle \pi} & & \downarrow{\scriptstyle \pi_{r,s}} \\
\mathbf{P}^r & \xleftarrow{p_1} & \mathbf{P}^r \times \mathbf{P}^s & \xrightarrow{p_2} & \mathbf{P}^s
\end{array}
$$

---

[8]We follow Grothendieck's notation: for a vector bundle $\mathscr{E}$, the projectivization $\mathbf{P}(\mathscr{E})$ is the space of *hyperplanes* in the fibers of $\mathscr{E}$.

[9]A line bundle on $\mathbf{P}^r \times \mathbf{P}^s$ is of type (a, b) if it is isomorphic to $p_1^* \mathscr{O}_{\mathbf{P}^r}(a) \otimes p_2^* \mathscr{O}_{\mathbf{P}^s}(b)$.

and

- the relative cone of ε_2 is the extremal ray $\mathbf{R}^+\ell$;

- the relative cone of π is the extremal ray $\mathbf{R}^+\ell'$;

- the relative cone of ε_1 is the extremal ray $\mathbf{R}^+\ell''$;

- the relative cone of $p_1 \circ \pi$ is the extremal face $\mathbf{R}^+\ell + \mathbf{R}^+\ell'$;

- the relative cone of $p_2 \circ \pi$ is the extremal face $\mathbf{R}^+\ell' + \mathbf{R}^+\ell''$.

What about the extremal face $\mathbf{R}^+\ell + \mathbf{R}^+\ell''$? It also corresponds to a morphism, which can be defined as follows.

The variety $Y_{r,s}$ contains a divisor isomorphic to $\mathbf{P}^r \times \mathbf{P}^s$ corresponding to the quotient $\mathscr{O}_{\mathbf{P}^s}(1)^{r+1}$ of $\mathscr{E}$. This divisor D belongs to the linear system $|\mathscr{O}_{Y_{r,s}}(1)|$, whose restriction to D has type $(1,1)$. There is an exact sequence

$$0 \to \mathbf{C} \xrightarrow{\cdot s_D} H^0(Y_{r,s}, \mathscr{O}_{Y_{r,s}}(1)) \to H^0(D, \mathscr{O}_D(1,1)) \to 0$$

which implies that the linear system $|\mathscr{O}_{Y_{r,s}}(1)|$ is base-point-free. It induces a morphism

$$c_{r,s} : Y_{r,s} \to \mathbf{P}^{(r+1)(s+1)}$$

which contracts P to a point and is an immersion on its complement. Its image $\widehat{Y}_{r\cdot s}$ is the cone over the Segre embedding of $\mathbf{P}^r \times \mathbf{P}^s$. It is therefore isomorphic to $\widehat{Y}_{s\cdot r}$. The relative cone of

$$c_{r,s} \circ \varepsilon_1 = c_{s,r} \circ \varepsilon_2 : X_{r\cdot s} \to \widehat{Y}_{r\cdot s}$$

is the extremal face $\mathbf{R}^+\ell + \mathbf{R}^+\ell''$. All contractions are displayed in the following nice hexagonal commutative diagram:

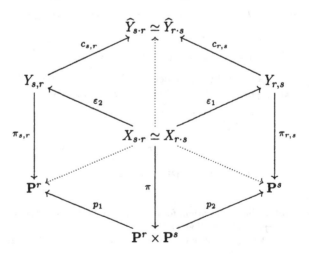

Solid arrows are contractions of extremal rays and dotted arrows are contractions of extremal 2-dimensional subcones.

1.37. Cubic surfaces. Let X be a smooth cubic surface in $\mathbf{P}^3$. It contains 27 lines $L_1, \ldots, L_{27}$, each one has self-intersection -1 and their classes span the 7-dimensional vector space $N_1(X)_{\mathbf{R}}$ ([H1], V, prop. 4.8 and th. 4.9). We will prove in Lemma 6.2(a) that each class $[L_i]$ spans an extremal ray in $\mathrm{NE}(X)$ and

$$\mathrm{NE}(X) = \overline{\mathrm{NE}}(X) = \sum_{i=1}^{27} \mathbf{R}^+[L_i] \subset \mathbf{R}^7$$

The cubic X is the image of the projective plane blown up at six points by the linear system of cubics passing through these points. The six exceptional divisors map onto six skew lines, say $L_1, \ldots, L_6$. Among the remaining 21 lines, 15 are the images of the strict transforms of the lines through two blown up points, and six are the images of the conics passing through all six points but one. If H is a hyperplane section of X, the classes of the lines on X are therefore

$$[L_1], \ldots, [L_6]$$
$$[H] - [L_i] - [L_j] \quad \text{for } 1 \leq i < j \leq 6$$
$$[2H] - [L_1] - \cdots - [L_6] + [L_i] \quad \text{for } 1 \leq i \leq 6$$

and the classes $[H], [L_1], \ldots, [L_6]$ form a basis for the real vector space $N_1(X)_{\mathbf{R}}$. One can see that even for surfaces, the geometry of the cone of curves can be quite intricate. And this is nothing compared to the next example.

1.38. A cone of curves that is not finitely generated. Again, we will not prove anything here and will discuss this example in a little more detail in 6.6. The blow-up $X \to \mathbf{P}^2$ at nine points contains infinitely many rational curves with self-intersection -1. As above, each such curve spans an extremal ray, and these rays are all distinct. The cone $\overline{\mathrm{NE}}(X)$ in the 10-dimensional vector space $N_1(X)_{\mathbf{R}}$ is not finitely generated.

1.10 Exceptional Locus of a Morphism

If X is a smooth variety, we denote its tangent bundle by T_X and a canonical divisor (any divisor whose associated line bundle is $(\det T_X)^*$) by K_X.

In what follows, X and Y are varieties.

1.39. Domain of definition of a rational map. Let $\pi : X \dashrightarrow Y$ be a rational map, defined over some dense open set $V \subset X$. Let X' be the closure in $X \times Y$ of the graph of $\pi|_V : V \to Y$. We will call it the graph of π. The first projection $p : X' \to X$ is birational. There is a largest open set U of X on which π is defined: it is the largest set over which p is an isomorphism.

If X is normal and Y is proper, p is proper and its fibers are connected by Zariski's theorem ([H1], III, cor. 11.4). If a fiber $p^{-1}(x)$ is a single point, x has a neighborhood V in X such that the map $p^{-1}(V) \to V$ induced by p is finite. Since it is birational and X is normal, it is an isomorphism (we made that argument in 1.13). It follows that $X - U$ is exactly the set of points of X that have positive-dimensional fibers, hence that $X - U$ *has codimension at least 2 in X.*

1.40. Exceptional locus of a birational morphism. Let $\pi : Y \to X$ be a birational morphism. The *exceptional locus* $\mathrm{Exc}(\pi)$ of π is the locus of points of Y where π is not a local isomorphism. We will denote it here by E.

If X is normal and π is proper, the fibers of π are connected by Zariski's theorem. If $y \in E$ and $\pi^{-1}(\pi(y))$ has at least two points, it has therefore positive dimension everywhere, hence is contained in E. This implies $E = \pi^{-1}(\pi(E))$ and $\pi(E)$ has codimension at least 2 in X. The largest open set over which $\pi^{-1} : X \dashrightarrow Y$ is defined is $X - \pi(E)$.

If X is normal and locally $\mathbf{Q}$-factorial[10] (e.g., smooth), *every irreducible component of E has codimension 1 in Y and its image has codimension at least 2 in X.* This can be seen as follows. Let $y \in E$ and $x = \pi(y)$. Identify the quotient fields $K(X)$ and $K(Y)$ by the isomorphism π^*, so that $\mathcal{O}_{X,x}$ is a proper subring of $\mathcal{O}_{Y,y}$. Let t be an element of $\mathfrak{m}_{Y,y}$ not in $\mathcal{O}_{X,x}$, and write its divisor as the difference of two effective divisors D' and D'' without common components. There exists a positive integer m such that mD' and mD'' are Cartier divisors, hence define elements u and v of $\mathcal{O}_{X,x}$ such that $t^m = \frac{u}{v}$. Both are actually in $\mathfrak{m}_{X,x} : v$ because t^m is not in $\mathcal{O}_{X,x}$ (otherwise, t would be since $\mathcal{O}_{X,x}$ is integrally closed), and $u = t^m v$ because it is in $\mathfrak{m}_{Y,y} \cap \mathcal{O}_{X,x} = \mathfrak{m}_{X,x}$. But the equations $u = v = 0$ define a subscheme Z containing x, which has codimension 2 in some neighborhood of x (it is the intersection of the codimension 1 subschemes mD' and mD''), whereas $\pi^{-1}(Z)$ is defined by $t^m v = v = 0$, hence by the sole equation $v = 0$: it has codimension 1 in Y, hence is contained in E. It follows that there is a codimension 1 component of E through every point of E, which proves what we wanted.

1.41. Ramification divisor of a morphism between smooth varieties. Assume that X and Y are smooth. To a morphism $\pi : Y \to X$ we associate its tangent map $T\pi : T_Y \to \pi^* T_X$. If X and Y have the same dimension n, its determinant induces a morphism $\wedge^n T_Y \to \pi^*(\wedge^n T_X)$ of invertible sheaves, hence a section of $\mathcal{O}_Y(K_Y - \pi^* K_X)$. If this section is not

---

[10]This means that any Weil divisor on X has a multiple that is a Cartier divisor.

identically zero (i.e., if π is generically étale[11]), it vanishes on an effective divisor $\mathrm{Ram}(\pi)$ called the *ramification divisor* of π. In particular,

$$K_Y \equiv \pi^* K_X + \mathrm{Ram}(\pi) \tag{1.11}$$

If π is moreover birational, the support of $\mathrm{Ram}(\pi)$ is the exceptional locus $\mathrm{Exc}(\pi)$ defined above.

1.42. Very ample divisors on a blow-up. Let X and Y be projective varieties. A birational morphism $\pi : Y \to X$ is the blow-up of a coherent sheaf of ideals $\mathscr{I}$ on X. The corresponding line bundle $\mathscr{O}_Y(1)$ is the inverse ideal sheaf $\pi^{-1} \mathscr{I} \cdot \mathscr{O}_Y$. It defines an effective divisor E on Y ([H1], II, th. 7.17 and prop. 7.13). If H is an ample divisor on X, the divisor $m\pi^* H - E$ is very ample on Y for all integers m sufficiently large by [H1], II, ex. 7.14(b).

When X is normal, the locus $\mathrm{Exc}(\pi)$ is contained in the support of E: by Zariski's theorem, any point in $\mathrm{Exc}(\pi)$ is on a curve C contracted by π, hence

$$0 < (m\pi^* H - E) \cdot C = -E \cdot C$$

which implies that C is contained in the support of E. It follows from [H1], II, ex. 7.11(c), that the support of E is exactly $\mathrm{Exc}(\pi)$ when X is smooth.

More generally, *if X is normal and locally $\mathbf{Q}$-factorial* (see footnote 10, p. 28), *there exists an effective $\mathbf{Q}$-Cartier $\mathbf{Q}$-divisor F on Y whose support is $\mathrm{Exc}(\pi)$ and such that $m\pi^* H - F$ is very ample for all integers m sufficiently large.*

This can be seen as follows: let H_Y be an ample divisor on Y. Since some multiple $r\pi(H_Y)$ of $\pi(H_Y)$ is a Cartier divisor, one can write $\pi^*(r\pi(H_Y)) = rH_Y + rF$, where F is a $\mathbf{Q}$-divisor with support contained in $\mathrm{Exc}(\pi)$. By 1.20, $mH - \pi(H_Y)$ is ample for all m sufficiently large, hence its inverse image $m\pi^* H - H_Y - F$ is nef and $m\pi^* H - F$ is ample by 1.22. Some multiple $m'(m\pi^* H - F)$ is very ample, and $m''\pi^* H - m'F$ is very ample for $m'' \geq mm'$ by 1.19.

It is worthwhile to stress the fact that when X is not locally $\mathbf{Q}$-factorial, the exceptional locus of π may well have codimension more than 1 (this will be a source of considerable trouble for us later on), in which case the support of E *cannot* be contained therein.

1.11 Rational Curves on Exceptional loci

Now that we have set up all the classical constructions involving curves and divisors on algebraic varieties, we will briefly discuss the problem of

---

[11]In characteristic zero, this property holds whenever π is surjective. In general, the extension $K(X) \subset K(Y)$ needs to be separable.

classifying algebraic varieties and explain how *rational* curves come into play. We will go back to this problem in much greater detail in Chapter 7.

The classification of algebraic varieties usually involves three steps:

- define when two varieties are equivalent;

- find a "simple" member in each equivalence class;

- explain how a given algebraic variety is related to the "simple" member(s) of its equivalence class.

The only equivalence relation that makes sense for the first step is *birational equivalence*: two varieties are birationally equivalent if there is a birational isomorphism between them.

For smooth projective curves, there is not much to say as to the last two steps, since each such curve is the only member of its birational equivalence class.

For smooth projective surfaces, the situation is much more interesting, but still completely understood (see [H1], V, §5). Let X be a smooth projective surface. Any rational curve with self-intersection -1 on X can be contracted by a birational morphism $X \to X'$, where X' is a smooth projective surface. Iterating this process, we end up with a birational *morphism* $X \to X_0$, where X_0 is a smooth surface with no such curve. It is called *minimal*. There are two cases:

- either X is *not* covered by rational curves, the minimal model X_0 is unique, and K_{X_0} is nef;

- or else X is covered by rational curves, the minimal model X_0 is *not* unique, and K_{X_0} is *not* nef.

Here is a simple illustration of the second case: starting with $\mathbf{P}^2$ blown up at two points, we may also blow down the strict transform of the line that joins them to get $\mathbf{P}^1 \times \mathbf{P}^1$. The surfaces $\mathbf{P}^2$ and $\mathbf{P}^1 \times \mathbf{P}^1$ are two distinct birational minimal surfaces.

Things get more complicated in higher dimensions, and there are distinct "minimal" models with nef canonical classes which are birationally isomorphic. However, we will prove two elementary results that show the importance of:

- rational curves;

- whether or not the canonical class is nef.

Before getting down to work, I will quote Kollár ([K2], p. 249):

> The more rational curves X contains, the more complicated the birational geometry of X.

Proposition 1.43 *Let X and Y be varieties, with X smooth, and let $\pi :$ $Y \to X$ be a proper birational morphism which is not an isomorphism. Through a general point of each component of* $\mathrm{Exc}(\pi)$, *there exists a rational curve contracted by* π.

The proof shows more precisely that (even if π is not proper) every irreducible component of the hypersurface $\mathrm{Exc}(\pi)$ is *ruled:* it is birational to a product $\mathbf{P}^1 \times Z$.

PROOF OF THE PROPOSITION. Set $E = \mathrm{Exc}(\pi)$. Upon replacing Y with its normalization, we may assume that Y is nonsingular in codimension 1. Recall also that each component of E has codimension 1 (1.40). Let y be a general point of a component of E. By shrinking Y, we may assume that E and Y are smooth irreducible. Again by shrinking X and Y, we may assume that $\overline{\pi(E)}$ is smooth, of codimension at least 2 (by 1.40 again). Let $\varepsilon_1 : X_1 \to X$ its blow-up. By the universal property of blow-ups ([H1], II, prop. 7.14), there is a factorization

$$\pi : Y \xrightarrow{\pi_1} X_1 \xrightarrow{\varepsilon_1} X$$

and $\overline{\pi_1(E)}$ is contained in the support of the exceptional divisor E_1 of ε_1. If the codimension of $\overline{\pi_1(E)}$ in X_1 is at least 2, the divisor E is contained in the exceptional locus of π_1 and we may repeat the construction to get a factorization

$$\pi : Y \xrightarrow{\pi_i} X_i \xrightarrow{\varepsilon_i} X_{i-1} \xrightarrow{\varepsilon_{i-1}} \cdots \xrightarrow{\varepsilon_2} X_1 \xrightarrow{\varepsilon_1} X$$

as long as the codimension of $\overline{\pi_{i-1}(E)}$ in X_{i-1} is at least 2. We have

$$
\begin{aligned}
K_{X_i} &= \varepsilon_i^* K_{X_{i-1}} + c_i E_i \\
&= (\varepsilon_1 \circ \cdots \circ \varepsilon_i)^* K_X + c_i E_i + \cdots + c_1 E_1
\end{aligned}
$$

where E_i is the exceptional divisor of ε_i and $c_i = \mathrm{codim}_{X_{i-1}}(\overline{\pi_{i-1}(E)}) - 1$ ([H1], II, ex. 8.5). Since π_i is birational, $\pi_i^* \mathcal{O}_{X_i}(K_{X_i})$ is a subsheaf of $\mathcal{O}_Y(K_Y)$. Moreover, since $\pi_i(E)$ is contained in the support of E_i, the divisor $\pi_i^* E_i - E$ is effective. It follows that $\pi^* \mathcal{O}_X(K_X) + (c_{i-1} + \cdots + c_0)E$ is a subsheaf of $\mathcal{O}_Y(K_Y)$. Since there are no infinite ascending sequences of subsheaves of a coherent sheaf on a Noetherian scheme, the process must terminate at some point: $\overline{\pi_i(E)}$ is a divisor in X_i for some i, hence E is not contained in the exceptional locus of π_i (by 1.40 again). The morphism π_i then induces a birational isomorphism between E and E_i, and the latter is ruled: more precisely, through every point of E_i there is a rational curve contracted by ε_i. This proves the proposition. $\qquad \square$

Corollary 1.44 *Let X and Y be varieties. Assume that X is smooth and that Y is proper and contains no rational curves. Any rational map $X \dashrightarrow Y$ is everywhere defined.*

PROOF. Let $X' \subset X \times Y$ be the graph of a rational map $\pi : X \dashrightarrow Y$ as defined in 1.39. The first projection induces a proper birational morphism $p : X' \to X$. Assume its exceptional locus $\mathrm{Exc}(p)$ is nonempty. By the proposition, there exists a rational curve on $\mathrm{Exc}(p)$ that is contracted by p. Since Y contains no rational curves, it must also be contracted by the second projection, which is absurd since it is contained in $X \times Y$. Hence $\mathrm{Exc}(p)$ is empty and π is defined everywhere.   □

Under the hypotheses of the proposition, one can say more if Y also is smooth.

Proposition 1.45 *Let X and Y be smooth projective varieties and let $\pi : Y \to X$ be a birational morphism that is not an isomorphism. There exists a rational curve C on Y contracted by π such that $K_Y \cdot C < 0$.*

PROOF. Let E be the exceptional locus of π. By 1.40, $\pi(E)$ has codimension at least 2 in X. Let x be a point of $\pi(E)$. By Bertini's theorem ([H1], II, th. 8.18), a general hyperplane section of X is smooth and connected, and a general hyperplane section of X *passing through x* has the same property.[12] It follows that by taking $(\dim(X) - 2)$ general hyperplane sections, we get a smooth surface S in X that meets $\pi(E)$ in a finite set containing x. Moreover, taking one more hyperplane section, we get on S a smooth curve C_0 that meets $\pi(E)$ only at x and a smooth curve C that does not meet $\pi(E)$.

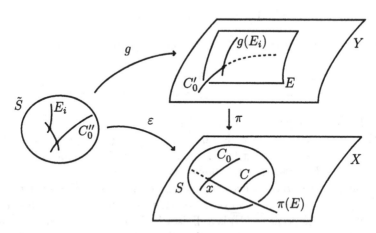

Construction of a rational curve $g(E_i)$ in the exceptional locus E of π.

---

[12]Let H be a very ample divisor on X. On the blow-up $\varepsilon : \tilde{X} \to X$ of x, with exceptional divisor F, there exists a positive integer m such that $m\varepsilon^* H - F$ is very ample (1.42), and Bertini's theorem implies that a general element of the corresponding linear system is smooth and connected. Its intersection with F is a hyperplane; hence, it projects onto a smooth hypersurface of X passing through x.

By construction,
$$K_X \cdot C = K_X \cdot C_0$$

Recall from (1.11) the formula $K_Y \equiv \pi^* K_X + \mathrm{Ram}(\pi)$, where the support of the divisor $\mathrm{Ram}(\pi)$ is exactly E. Since the curve $C' = \pi^{-1}(C)$ does not meet E, we have
$$K_Y \cdot C' = K_X \cdot C$$

by the projection formula (1.5). On the other hand, since the strict transform
$$C_0' = \overline{\pi^{-1}(C_0 - \pi(E))}$$

of C_0 does meet E, we have
$$K_Y \cdot C_0' = (\pi^* K_X + \mathrm{Ram}(\pi)) \cdot C_0' > (\pi^* K_X) \cdot C_0' = K_X \cdot C_0$$

hence
$$K_Y \cdot C_0' > K_Y \cdot C' \qquad (1.12)$$

The rational map $\pi^{-1} : S \dashrightarrow Y$ might not be a morphism, but its indeterminacies can be resolved by blowing up a finite number of points[13] of $S \cap \pi(E)$ to get a morphism
$$g : \tilde{S} \xrightarrow{\varepsilon} S \xdashrightarrow{\pi^{-1}} Y$$

whose image is the strict transform of S. The curve $C'' = \varepsilon^* C$ is irreducible and $g_* C'' = C'$. For C_0, we write
$$\varepsilon^* C_0 = C_0'' + \sum_i m_i E_i$$

where the m_i are nonnegative integers, the E_i are exceptional divisors for ε (hence in particular rational curves), and $g_* C_0'' = C_0'$. Since C and C_0 are linearly equivalent on S, we have
$$C'' \equiv C_0'' + \sum_i m_i E_i$$

on $\tilde{S}$ hence, by applying g_*,
$$C' \equiv C_0' + \sum_i m_i (g_* E_i)$$

Taking intersections with K_Y, we get
$$K_Y \cdot C' \equiv K_Y \cdot C_0' + \sum_i m_i (K_Y \cdot g_* E_i)$$

---

[13]This is because a birational morphism between smooth projective surfaces is a composition of blow-ups ([H1], V, cor. 5.4): apply it to the projection from the graph of π^{-1} to Y.

It follows from (1.12) that $(K_Y \cdot g_* E_i)$ is negative for some i. In particular, $g(E_i)$ is not a point, hence is a rational curve on Y. Moreover, $\pi(g(E_i)) = \varepsilon(E_i)$; hence $g(E_i)$ is contracted by π. $\square$

We will close this section and chapter by interpreting the last results from the point of view of the classification of algebraic varieties discussed earlier. Let $\mathscr{C}$ be a birational equivalence class of smooth projective varieties. One aims at finding a "simplest" member in $\mathscr{C}$. The picture we now have is the following.

Proposition 1.44 says that if $\mathscr{C}$ contains a (smooth) variety X_0 without rational curves, it is automatically "minimal" in the sense that for each member X of $\mathscr{C}$, there is a birational *morphism* $X \to X_0$. Moreover, by Proposition 1.45, X_0 is uniquely determined up to isomorphism (this is the best possible situation!).

If $\mathscr{C}$ contains a variety X_1 such that K_{X_1} has nonnegative degree on every rational curve[14] (so this is weaker than containing no rational curves), one may only say by Proposition 1.45 that there is no birational morphism from X_1 to another member of $\mathscr{C}$. For surfaces, it is true that X_1 is still "minimal" in the above sense, but this is not true in higher dimensions as shown by the following example.

Example 1.46 Let r be a positive integer and let $X = X_{r \cdot r}$ and $Y = Y_{r,r}$ be the smooth $(2r+1)$-dimensional projective varieties constructed in 1.36 (we drop the indices for simplicity). The subvariety $\widehat{Y} = \widehat{Y}_{r \cdot r}$ of $\mathbf{P}^{(r+1)^2}$ is the cone over the Segre embedding of $\mathbf{P}^r \times \mathbf{P}^r$, hence has an involution σ that interchanges the two rulings. This involution lifts to X, which is the blow-up of the vertex of $\widehat{Y}$. There is a commutative diagram

$$
\begin{array}{ccc}
X & \xrightarrow{\ \sigma\ } & X \\
{\scriptstyle \varepsilon}\big\downarrow & & \big\downarrow{\scriptstyle \varepsilon} \\
Y & \dashrightarrow{\ \alpha\ } & Y \\
{\scriptstyle c}\big\downarrow & & \big\downarrow{\scriptstyle c} \\
\widehat{Y} & \xrightarrow{\ \sigma\ } & \widehat{Y}
\end{array}
$$

where $c = \varphi_{\mathcal{O}_Y(1)}$. The rational map

$$\alpha = \varepsilon \circ \sigma \circ \varepsilon^{-1} : Y \dashrightarrow Y$$

is not a morphism. However, K_Y does have negative degree on the class ℓ' of a line in a fiber of $Y \to \mathbf{P}^r$, so we do not have yet our counterexample. Indeed, $\mathcal{O}_Y(K_Y) = \mathcal{O}_Y(-(r+2))$; hence $K_Y \cdot \ell' = -(r+2)$. This is easy

---

[14]We will show later in Theorem 6.1 that this is equivalent to K_{X_1} nef.

to fix: consider the smooth hypersurface H cut out on $\widehat{Y}$ by a general hypersurface of degree $2(r + 3)$ in $\mathbf{P}^{(r+1)^2}$, the double cover $\pi' : Y' \to Y$ branched along $c^{-1}(H)$, and the double cover $\pi'' : Y'' \to Y$ branched along $c^{-1}(\sigma(H))$. We have

$$\mathcal{O}_{Y'}(K_{Y'}) \simeq \pi'^* \mathcal{O}_Y(1) \simeq (c \circ \pi')^* \mathcal{O}_{\mathbf{P}^{(r+1)^2}}(1)$$

and similarly for $\mathcal{O}_{Y''}(K_{Y''})$, hence $K_{Y'}$ and $K_{Y''}$ are nef, but the birational map $Y' \dashrightarrow Y''$ induced by α is *not* a morphism.

1.47. Mori's program. From this point of view, Mori's program can be seen as an algorithm that, starting from a given smooth projective variety, explains how to obtain, if possible in the same birational equivalence class, a variety that has nef canonical class, so that it is "minimal" in the above sense. The steps of this algorithm will be related to contractions of extremal rays, as described in 1.16. Although much progress has been made in the last twenty years, this program is at present only known to work in full generality in dimension at most 3.

Many technical complications will arise along the way: we will be forced to allow singular varieties, and the picture is actually far more complicated than what I have described here. It is not until Chapter 7, after having accumulated considerable technical expertise, that we will be able to go back to this question and make any kind of significant progress.

1.12 Exercises

1. Let **k** be a field and let U be the complement of the origin in the affine plane $\mathbf{A}_{\mathbf{k}}^2$. Compute $H^1(U, \mathcal{O}_U)$ and deduce that U is not affine.

2. Let **k** be a field and, in the affine space $\mathbf{A}_{\mathbf{k}}^4$, let P_1 be the plane defined by $T_1 = T_2 = 0$ and let P_2 be the plane defined by $T_3 = T_4 = 0$. Let U be the complement in $\mathbf{A}_{\mathbf{k}}^4$ of $P_1 \cup P_2$.

   Compute $H^2(U, \mathcal{O}_U)$ and deduce that the codimension 2 subvariety $P_1 \cup P_2$ of $\mathbf{A}_{\mathbf{k}}^4$ is not the set-theoretic intersection of two hypersurfaces.

3. Let X be a projective scheme, let $\mathscr{F}$ be a coherent sheaf on X, and let $H_1, \ldots, H_r$ be ample divisors on X. Prove that the set

   $$\{(m_1, \ldots, m_r) \in \mathbf{N}^r \mid \exists i > 0 \quad H^i(X, \mathscr{F}(m_1 H_1 + \cdots + m_r H_r)) \neq 0\}$$

   is finite.

4. Let X be a proper scheme that is a finite union of (closed) sub-schemes $X_1, \ldots, X_s$ (e.g., its irreducible components). Let $D_1, \ldots, D_r$

be Cartier divisors on X, with $r \geq \dim(X)$. Prove the following equality

$$D_1 \cdots D_r = \sum_{i=1}^{s} D_1|_{X_i} \cdots D_r|_{X_i}$$

5. Let X be a projective scheme. Recall from 1.30 that a Cartier divisor D on X is *big* if

$$\liminf_{m \to +\infty} \frac{h^0(X, mD)}{m^n} > 0$$

(we do not assume that D is nef). Show that the following properties are equivalent:

(i) D is big;

(ii) D is the sum of an ample and of an effective divisor;

(iii) there exists a positive integer m such that the rational map

$$X \dashrightarrow \mathbf{P}H^0(X, mD)$$

associated with the linear system $|mD|$ is birational onto its image.

6. Let $D_1, \ldots, D_n$ be nef divisors on a proper variety X of dimension n. Prove

$$D_1 \cdots D_n \geq (D_1^n)^{1/n} \cdots (D_n^n)^{1/n}$$

(*Hint:* First do the case when the divisors are ample by induction on n, using the Hodge index theorem when $n = 2$ ([H1], V, rem. 1.9.1).)

7. Let X be an abelian variety of dimension n and let D and H be divisors on X, with H ample. Show that D is nef (resp. ample) if and only if, for all $1 \leq i \leq n$, we have

$$D^i \cdot H^{n-i} \geq 0 \qquad (\text{resp. } D^i \cdot H^{n-i} > 0)$$

(*Hint:* Use Theorem 2 in [MuK]: the roots of the polynomial

$$P(m) = \chi(X, mH + D) = \frac{1}{n!}(mH + D)^n$$

are all real and $H^i(X, D) = 0$ for i greater than the number of non-negative roots of P.)

2

Parametrizing Morphisms

We continue in this chapter our exposition of background results with a much more specific purpose. We will concentrate on basically one object, whose construction dates back to Grothendieck in 1962: the space parametrizing curves on a given variety, or more precisely morphisms from a given smooth projective curve C to a given smooth quasi-projective variety. Mori's techniques, which will be discussed in the next chapter, make systematic use of these spaces in a rather exotic way.

We will not reproduce Grothendieck's construction, since it is very nicely explained in [Gr5] and only the end product will be important for us. However, we will explain in some detail in what sense these spaces are *parameter spaces,* and work out their local structure. Roughly speaking, as in many deformation problems, the tangent space to such a parameter space at a point is $H^0(C, \mathscr{F})$, where $\mathscr{F}$ is some vector bundle on C, first-order deformations are obstructed by elements of $H^1(C, \mathscr{F})$, and the dimension of the parameter space is therefore bounded from below by the difference $h^0(C, \mathscr{F}) - h^1(C, \mathscr{F})$. The crucial point is that since C has dimension 1, this difference is the Euler characteristic of $\mathscr{F}$, which can be computed from numerical data by the Riemann–Roch theorem.

The proofs are written only for the simplest version of these spaces of morphisms. All kinds of additional structures (see Section 2.3) will be needed later on. Since the proofs are usually similar, not very instructive, and would take us too far, we limit ourselves to a list of the properties that will be needed later on, without proofs.

To familiarize the reader with these spaces, we study in the last section the scheme of lines contained in a subvariety of a projective space (which is a nonparametrized version of the spaces of morphisms discussed above), showing that even in harmless-looking situations, they can be everywhere nonreduced.

We give in the exercises at the end of this chapter a rather complete description of this scheme of lines for the *Fermat hypersurface* X_N^d defined by the equation

$$x_0^d + \cdots + x_N^d = 0$$

in $\mathbf{P}_\mathbf{k}^N$. It turns out that when $n \leq d < \mathrm{char}(\mathbf{k})$, all lines contained in X_N^d can be explicitly described (thereby extending results from [AK]). The corresponding scheme of lines exhibits quite interesting behavior. The case where $d - 1$ is a power of the characteristic is also worth playing around with. We prove that the Fermat hypersurface is in this case unirational.[1]

2.1   Parametrizing Rational Curves

Let $\mathbf{k}$ be a field. Any $\mathbf{k}$-morphism f from $\mathbf{P}_\mathbf{k}^1$ to $\mathbf{P}_\mathbf{k}^N$ can be written as

$$f(u, v) = (F_0(u, v), \ldots, F_N(u, v)) \tag{2.1}$$

where $F_0, \ldots, F_N$ are homogeneous polynomials in two variables, of the same degree d, with no nonconstant common factor in $\mathbf{k}[U, V]$. We are going to show that there exist universal integral polynomials in the coefficients of the F_i that vanish if and only they have a nonconstant common factor in $\mathbf{k}[U, V]$. The F_i have no nonconstant common factor in $\mathbf{k}[U, V]$ if and only if they have no nontrivial common zero in an algebraic closure $\bar{\mathbf{k}}$ of $\mathbf{k}$. By the Nullstellensatz, this holds if and only if the ideal generated by $F_0, \ldots, F_N$ in $\bar{\mathbf{k}}[U, V]$ contains some power of the maximal ideal (U, V). This in turn means that for some m, the map

$$\begin{array}{ccc} (\bar{\mathbf{k}}[U, V]_{m-d})^{N+1} & \to & \bar{\mathbf{k}}[U, V]_m \\ (G_0, \ldots, G_N) & \mapsto & \sum_{i=0}^N F_i G_i \end{array}$$

is surjective (here $\mathbf{k}[U, V]_m$ is the vector space of homogeneous polynomials of degree m). This map being linear and defined over $\mathbf{k}$, we conclude that $F_0, \ldots, F_N$ have a nonconstant common factor in $\mathbf{k}[U, V]$ if and only if, for all m, all $(m + 1)$-minors of some universal matrix whose entries are linear integral combinations of the coefficients of the F_i vanish.

Therefore, morphisms of degree d from $\mathbf{P}^1$ to $\mathbf{P}^N$ are parametrized[2] by a Zariski open set of the projective space $\mathbf{P}((S^d \mathbf{k}^2)^{N+1})$. We denote

---

[1]This was proved for N odd in [SK].

[2]One should stress that we are here parametrizing *maps* from $\mathbf{P}^1$ to $\mathbf{P}^N$, not their images. For example, for $d = 1$, the parameter space is not the Grassmannian of lines in $\mathbf{P}^N$. The latter is the quotient of $\mathrm{Mor}_1(\mathbf{P}^1, \mathbf{P}^N)$ by the action of the group $\mathrm{PGL}(2, \mathbf{k})$ of automorphisms of $\mathbf{P}^1$.

this quasi-projective variety $\mathrm{Mor}_d(\mathbf{P}^1, \mathbf{P}^N)$. Note that these morphisms fit together into a *universal morphism*

$$f^{\mathrm{univ}} : \quad \begin{array}{ccc} \mathbf{P}^1 \times \mathrm{Mor}_d(\mathbf{P}^1, \mathbf{P}^N) & \to & \mathbf{P}^N \times \mathrm{Mor}_d(\mathbf{P}^1, \mathbf{P}^N) \\ ((u,v), f) & \mapsto & ((F_0(u,v), \ldots, F_N(u,v)), f) \end{array}$$

Finally, morphisms from $\mathbf{P}^1$ to $\mathbf{P}^N$ are parametrized by the (locally Noetherian) disjoint union

$$\mathrm{Mor}(\mathbf{P}^1, \mathbf{P}^N) = \bigsqcup_{d \geq 0} \mathrm{Mor}_d(\mathbf{P}^1, \mathbf{P}^N)$$

of quasi-projective varieties.

Let now X be a subscheme of $\mathbf{P}^N$ defined by homogeneous equations $G_1, \ldots, G_m$. Morphisms of degree d from $\mathbf{P}^1$ to X are parametrized by the subscheme $\mathrm{Mor}_d(\mathbf{P}^1, X)$ of $\mathrm{Mor}_d(\mathbf{P}^1, \mathbf{P}^N)$ defined by the equations

$$G_j(F_0, \ldots, F_N) = 0 \qquad \text{for } j = 1, \ldots, m$$

Again, morphisms from $\mathbf{P}^1$ to X are parametrized by the disjoint union

$$\mathrm{Mor}(\mathbf{P}^1, X) = \bigsqcup_{d \geq 0} \mathrm{Mor}_d(\mathbf{P}^1, X)$$

of quasi-projective schemes.

If now X can be defined by homogeneous equations $G_1, \ldots, G_m$ with coefficients in a subring R of $\mathbf{k}$, the scheme $\mathrm{Mor}_d(\mathbf{P}^1, X)$ has the same property. If $\mathfrak{m}$ is a maximal ideal of R, one may consider the reduction $X_{\mathfrak{m}}$ of X modulo $\mathfrak{m}$: this is the subscheme of $\mathbf{P}^N_{R/\mathfrak{m}}$ defined by the reductions of the G_j modulo $\mathfrak{m}$. Because the equations defining the complement of $\mathrm{Mor}_d(\mathbf{P}^1, \mathbf{P}^N)$ in $\mathbf{P}((S^d\mathbf{k}^2)^{N+1})$ are the same for all fields, $\mathrm{Mor}_d(\mathbf{P}^1, X_{\mathfrak{m}})$ is the reduction of $\mathrm{Mor}_d(\mathbf{P}^1, X)$ modulo $\mathfrak{m}$. In fancy terms, one may express this as follows: if $\mathscr{X}$ is a scheme over $\mathrm{Spec}\, R$, the R-morphisms $\mathbf{P}^1_R \to \mathscr{X}$ are parametrized by (the R-points of) a locally Noetherian scheme

$$\mathrm{Mor}(\mathbf{P}^1_R, \mathscr{X}) \to \mathrm{Spec}\, R$$

and the fiber of a closed point $\mathfrak{m}$ is the space $\mathrm{Mor}(\mathbf{P}^1_{R/\mathfrak{m}}, \mathscr{X}_{\mathfrak{m}})$.

2.2 Parametrizing Morphisms

2.1. The space $\mathrm{Mor}(Y, X)$. Grothendieck vastly generalized the preceding construction: if X and Y are varieties defined over a field $\mathbf{k}$, with X quasi-projective and Y projective, he shows ([Gr5], 4(c)) that morphisms from Y to X are parametrized by a locally Noetherian scheme $\mathrm{Mor}(Y, X)$. As we saw in the case $Y = \mathbf{P}^1$ and $X = \mathbf{P}^N$, this scheme will in general have countably many components. One way to remedy that is to fix an ample divisor H on X and a polynomial P: the subscheme $\text{Mor}_P(Y, X)$ of $\text{Mor}(Y, X)$ that parametrizes morphisms $f : Y \to X$ with fixed *Hilbert polynomial*

$$P(m) = \chi(Y, mf^*H)$$

is now quasi-projective over $\mathbf{k}$, and $\text{Mor}(Y, X)$ is the disjoint union of the $\text{Mor}_P(Y, X)$, for all polynomials P. Note that when Y is a curve, fixing the Hilbert polynomial amounts to fixing the degree of the 1-cycle f_*Y for the embedding of X defined by some multiple of H.

Let us make more precise this notion of *parameter space*. First, there is as above a universal morphism

$$f^{\text{univ}} : Y \times \text{Mor}(Y, X) \to X \times \text{Mor}(Y, X)$$

such that, for each point t of $\text{Mor}(Y, X)$, the morphism $f_t^{\text{univ}} : Y \to X$ is the morphism corresponding to t (conversely, for any morphism $f : Y \to X$, we will write $[f]$ for the corresponding point of $\text{Mor}(Y, X)$).

The second property, called the *universal* property, is more subtle: it says that for any family of morphisms $f_t : Y \to X$ parametrized by a scheme T, the map $T \to \text{Mor}(Y, X)$ obtained by sending t to $[f_t]$ is algebraic. The correct way to express that is to say that there is a one-to-one correspondence between

- morphisms $\varphi : T \to \text{Mor}(Y, X)$ and

- T-morphisms $f : Y \times T \to X \times T$

obtained by sending φ to the pull-back

$$f(y, t) = \left(pr_1 \circ f^{\text{univ}}(y, \varphi(t)), t\right)$$

of f^{univ}. Given f (or φ), we will call the morphism

$$
\begin{array}{cccc}
\text{ev} : & Y \times T & \to & X \\
& (y, t) & \mapsto & f_t(y) = \varphi(t)(y)
\end{array}
$$

the *evaluation map*.

Examples 2.2 (1) When Y is a point $\{\star\}$, morphisms to X are just points of X. They are therefore (quite tautologically) parametrized by X itself, the universal morphism being

$$
\begin{array}{cccc}
f^{\text{univ}} : & \{\star\} \times X & \to & X \times X \\
& (\star, x) & \mapsto & (x, x)
\end{array}
$$

More generally, when Y is a reduced scheme of dimension 0, the scheme $\text{Mor}(Y, X)$ is isomorphic to X^Y.

(2) When $Y = \operatorname{Spec} k[\varepsilon]/(\varepsilon^2)$, a map $Y \to X$ corresponds to the data of a point x of X and a tangent vector in $T_{X,x}$. When X is smooth, $\operatorname{Mor}(\operatorname{Spec} k[\varepsilon]/(\varepsilon^2), X)$ is the total space of the tangent bundle to X.

(3) Let d be an integer ≥ 4. If k is algebraically closed and of characteristic zero or at least $d + 1$ and X is the (smooth) Fermat threefold defined by the equation

$$x_0^d + x_1^d + x_2^d + x_3^d + x_4^d = 0$$

in $\mathbf{P}_k^4$, we will show in 2.16 that for $d \geq 4$, the scheme $\operatorname{Mor}_1(\mathbf{P}^1, X)$, which parametrizes rational curves on X whose image is a line, is nowhere reduced.

2.3. The tangent space to $\operatorname{Mor}(Y, X)$. We will use the universal property to determine the Zariski tangent space to $\operatorname{Mor}(Y, X)$ at a point $[f]$. This vector space parametrizes by definition morphisms from $T = \operatorname{Spec} k[\varepsilon]/(\varepsilon^2)$ to $\operatorname{Mor}(Y, X)$ with image $[f]$ ([H1], II, ex. 2.8), hence extensions of f to T-morphisms

$$f_\varepsilon : Y \times \operatorname{Spec} k[\varepsilon]/(\varepsilon^2) \to X \times \operatorname{Spec} k[\varepsilon]/(\varepsilon^2)$$

which should be thought of as first-order infinitesimal deformations of f.

Proposition 2.4 *Let X and Y be varieties, with X quasi-projective and Y projective, and let $f : Y \to X$ be a morphism. One has*

$$T_{\operatorname{Mor}(Y,X),[f]} \simeq H^0(Y, \mathscr{H}om(f^*\Omega_X^1, \mathscr{O}_Y))$$

PROOF. Assume first that Y and X are affine and write $Y = \operatorname{Spec}(B)$ and $X = \operatorname{Spec}(A)$ (where A and B are finitely generated k-algebras). Let $f^\sharp : A \to B$ be the morphism corresponding to f. We are looking for $(k[\varepsilon]/(\varepsilon^2))$-algebra homomorphisms $f_\varepsilon^\sharp : A[\varepsilon] \to B[\varepsilon]$ of the type

$$f_\varepsilon^\sharp(a) = f(a) + \varepsilon g(a)$$

where a is in A. The equality $f_\varepsilon^\sharp(aa') = f_\varepsilon^\sharp(a) f_\varepsilon^\sharp(a')$ is equivalent to

$$g(aa') = f^\sharp(a)g(a') + f^\sharp(a')g(a)$$

In other words, $g : A \to B$ must be a k-derivation of the A-module B, hence must factor as $g : A \to \Omega_A \to B$ ([H1], II, §8). Such extensions are therefore parametrized by $\operatorname{Hom}_A(\Omega_A, B) = \operatorname{Hom}_B(\Omega_A \otimes_A B, B)$.

In general, cover X by affine open subsets $U_i = \operatorname{Spec}(A_i)$ and Y by affine open subsets $V_i = \operatorname{Spec}(B_i)$ such that $f(V_i)$ is contained in U_i. First-order extensions of $f|_{V_i}$ are parametrized by $h_i \in \operatorname{Hom}_{B_i}(\Omega_{A_i} \otimes_{A_i} B_i, B_i) = H^0(V_i, \mathscr{H}om(f^*\Omega_X^1, \mathscr{O}_Y))$. To glue these, we need the compatibility condition

$$h_i|_{V_i \cap V_j} = h_j|_{V_i \cap V_j}$$

which is exactly saying that the h_i define a global section on Y.   □

In particular, when X is smooth along the image of f,

$$T_{\text{Mor}(Y,X),[f]} \simeq H^0(Y, f^*T_X)$$

When Y is smooth, it follows that $H^0(Y, T_Y)$ is the tangent space at the identity to the group of automorphisms of Y. The image of the canonical morphism $H^0(Y, T_Y) \to H^0(Y, f^*T_X)$ corresponds to the deformations of f by reparametrizations.

2.5. The local structure of $\text{Mor}(Y, X)$. We prove the result mentioned in the introduction of this chapter. Its main use will be to provide a lower bound for the dimension of $\text{Mor}(Y, X)$ at a point $[f]$, thereby allowing us in certain situations to produce many deformations of f. This lower bound is very accessible, via the Riemann–Roch theorem, when Y is a curve (see 2.11).

Theorem 2.6 *Let X and Y be projective varieties and let $f : Y \to X$ be a morphism such that X is smooth along $f(Y)$. Locally around $[f]$, the scheme $\text{Mor}(Y, X)$ can be defined by $h^1(Y, f^*T_X)$ equations in a nonsingular variety of dimension $h^0(Y, f^*T_X)$. In particular, any irreducible component of $\text{Mor}(Y, X)$ through $[f]$ has dimension at least*

$$h^0(Y, f^*T_X) - h^1(Y, f^*T_X)$$

PROOF. Locally around the **k**-point $[f]$, the **k**-scheme $\text{Mor}(Y, X)$ can be defined by certain polynomial equations $P_1, \ldots, P_m$ in an affine space $\mathbf{A}_k^n$. The rank r of the corresponding Jacobian matrix $((\partial P_i/\partial x_j)([f]))$ is the codimension of the Zariski tangent space $T_{\text{Mor}(Y,X),[f]}$ in $\mathbf{k}^n$. The subvariety V of $\mathbf{A}_k^n$ defined by r equations among the P_i for which the corresponding rows have rank r is smooth at $[f]$ with the same Zariski tangent space as $\text{Mor}(Y, X)$.

Letting $h^i = h^i(Y, f^*T_X)$, we are going to show that $\text{Mor}(Y, X)$ can be locally around $[f]$ defined by h^1 equations inside the smooth h^0-dimensional variety V. For that, it is enough to show that in the regular local ring $R = \mathcal{O}_{V,[f]}$, the ideal I of functions vanishing on $\text{Mor}(Y, X)$ can be generated by h^1 elements. Note that since the Zariski tangent spaces are the same, I is contained in the square of the maximal ideal $\mathfrak{m}$ of R. Finally, by Nakayama's lemma ([Ma], th. 2.3), it is enough to show that the **k**-vector space $I/\mathfrak{m}I$ has dimension at most h^1.

The canonical morphism $\text{Spec}(R/I) \to \text{Mor}(Y, X)$ corresponds to an extension $f_{R/I} : Y \times \text{Spec}(R/I) \to X \times \text{Spec}(R/I)$ of f. Since $I^2 \subset \mathfrak{m}I$, the obstruction to extending it to a morphism $f_{R/\mathfrak{m}I} : Y \times \text{Spec}(R/\mathfrak{m}I) \to X \times \text{Spec}(R/\mathfrak{m}I)$ lies by Lemma 2.7 below in

$$H^1(Y, f^*T_X) \otimes_{\mathbf{k}} (I/\mathfrak{m}I)$$

Write this obstruction as

$$\sum_{i=1}^{h^1} a_i \otimes \bar{b}_i$$

where $(a_1, \ldots, a_{h^1})$ is a basis for $H^1(Y, f^*T_X)$ and $b_1, \ldots, b_{h^1}$ are in I. The obstruction vanishes modulo the ideal $(b_1, \ldots, b_{h^1})$, which means that the morphism $\mathrm{Spec}(R/I) \to \mathrm{Mor}(Y, X)$ lifts to a morphism

$$\mathrm{Spec}\big(R/\big(\mathfrak{m}I + (b_1, \ldots, b_{h^1})\big)\big) \to \mathrm{Mor}(Y, X)$$

In other words, the identity $R/I \to R/I$ factors as

$$R/I \to R/\big(\mathfrak{m}I + (b_1, \ldots, b_{h^1})\big) \xrightarrow{\pi} R/I$$

where π is the canonical projection. By Lemma 2.8 below,

$$I = \mathfrak{m}I + (b_1, \ldots, b_{h^1})$$

which means that $I/\mathfrak{m}I$ is generated by the classes of $b_1, \ldots, b_{h^1}$. $\square$

We now prove the two lemmas used in the proof above.

Lemma 2.7 *Let R be a finitely generated local $\mathbf{k}$-algebra with maximal ideal $\mathfrak{m}$ and residue field $\mathbf{k}$ and let I be an ideal contained in $\mathfrak{m}$ such that $\mathfrak{m}I = 0$. Let $f : Y \to X$ be a morphism and let*

$$f_{R/I} : Y \times \mathrm{Spec}(R/I) \to X \times \mathrm{Spec}(R/I)$$

be an extension of f. Assume X is smooth along the image of f. The obstruction to extending $f_{R/I}$ to a morphism

$$f_R : Y \times \mathrm{Spec}(R) \to X \times \mathrm{Spec}(R)$$

lies in

$$H^1(Y, f^*T_X) \otimes_{\mathbf{k}} I$$

PROOF. In the case where Y and X are affine, and with the notation of the proof of Proposition 2.4, we are looking for R-algebra liftings $f_R^{\sharp}$ fitting into the diagram

$$
\begin{array}{ccc}
A \otimes_{\mathbf{k}} R & \xdashrightarrow{f_R^{\sharp}} & B \otimes_{\mathbf{k}} R \\
\downarrow & & \downarrow \\
A \otimes_{\mathbf{k}} R/I & \xrightarrow{f_{R/I}^{\sharp}} & B \otimes_{\mathbf{k}} R/I
\end{array}
$$

Because X is smooth along the image of f and $I^2 = 0$, such a lifting exists ([H1], II, ex. 8.6), and two liftings differ by an R-derivation of $A \otimes_{\mathbf{k}} R$

into $B \otimes_k I$, that is, by an element of $\mathrm{Hom}_{A \otimes_k R}(\Omega_{A \otimes_k R/R}, B \otimes_k I)$. Since $\Omega_{A \otimes_k R/R} \simeq \Omega_A \otimes_k R$ ([H1], II, prop. 8.2A), we have

$$\mathrm{Hom}_{A \otimes_k R}(\Omega_{A \otimes_k R/R}, B \otimes_k I) \simeq \mathrm{Hom}_{A \otimes_k R}(\Omega_A \otimes_k R, B \otimes_k I)$$

Since $\mathfrak{m}I = 0$, we further have[3]

$$\begin{aligned}
\mathrm{Hom}_{A \otimes_k R}(\Omega_A \otimes_k R, B \otimes_k I) &\simeq \mathrm{Hom}_{A \otimes_k (R/\mathfrak{m})}(\Omega_A \otimes_k (R/\mathfrak{m}), B \otimes_k I) \\
&\simeq \mathrm{Hom}_A(\Omega_A, B \otimes_k I)
\end{aligned}$$

because $R/\mathfrak{m} = k$. In the end, we get

$$\begin{aligned}
\mathrm{Hom}_{A \otimes_k R}(\Omega_{A \otimes_k R/R}, B \otimes_k I) &\simeq \mathrm{Hom}_A(\Omega_A, B \otimes_k I) \\
&\simeq \mathrm{Hom}_B(B \otimes_k \Omega_A, B \otimes_k I) \\
&\simeq H^0(Y, \mathscr{H}om(f^*\Omega^1_X, \mathscr{O}_Y)) \otimes_k I \\
&\simeq H^0(Y, f^*T_X) \otimes_k I
\end{aligned}$$

To pass to the global case, one needs to patch up various local extensions to get a global one. There is an obstruction to doing that: on each intersection $V_i \cap V_j$, two extensions differ by an element of $H^0(V_i \cap V_j, f^*T_X) \otimes_k I$. These elements define a 1-cocycle, hence an element in $H^1(Y, f^*T_X) \otimes_k I$ whose vanishing is necessary and sufficient for a global extension to exist. In case such an extension exists, two extensions differ as above by an element of $H^0(Y, f^*T_X) \otimes_k I$. $\square$

Lemma 2.8 *Let R be a Noetherian local ring with maximal ideal $\mathfrak{m}$ and let I be an ideal in R contained in $\mathfrak{m}^2$. If the canonical projection $\pi : R \to R/I$ has a section, $I = 0$.*

PROOF. Let σ be a section of π. If a and b are in $\mathfrak{m}$, we have $\sigma \circ \pi(a) = a + a'$ and $\sigma \circ \pi(b) = b + b'$, where a' and b' are in I. Then

$$(\sigma \circ \pi)(ab) = (\sigma \circ \pi)(a)\,(\sigma \circ \pi)(b) = (a + a')(b + b') \in ab + \mathfrak{m}I + I^2$$

Since I is contained in $\mathfrak{m}^2$, we get, for any x in I,

$$0 = \sigma \circ \pi(x) \in x + \mathfrak{m}I$$

hence $I \subset \mathfrak{m}I$. Nakayama's lemma ([Ma], th. 2.2) implies $I = 0$. $\square$

---

[3]If A is a ring, J is an ideal in A, and M and N are A-modules such that $JN = 0$, the canonical map

$$\mathrm{Hom}_{A/J}(M/JM, N) \to \mathrm{Hom}_A(M, N)$$

is bijective.

2.3 Parametrizing Morphisms with Extra Structures

2.9. Morphisms with fixed points. We will need a slightly more general situation: fix a subscheme B of Y and a morphism $g : B \to X$. We want to study morphisms $f : Y \to X$ that restrict to g on B. Such morphisms can be parametrized by the fiber of $[g]$ under the restriction $\rho : \mathrm{Mor}(Y, X) \to \mathrm{Mor}(B, X)$. We denote this space by $\mathrm{Mor}(Y, X; g)$. At a point $[f]$ such that X is smooth along $f(Y)$, the tangent map to ρ is the restriction

$$H^0(Y, f^*T_X) \to H^0(B, g^*T_X)$$

hence *the tangent space to* $\mathrm{Mor}(Y, X; g)$ *is its kernel* $H^0(Y, f^*T_X \otimes I_B)$, where I_B is the ideal sheaf of B in Y.

Note that when B is finite, the scheme $\mathrm{Mor}(B, X)$ is smooth of dimension $\mathrm{lg}(B) \dim(X)$ by Theorem 2.6. It follows that $\mathrm{Mor}(Y, X; g)$ can be defined by at most $\mathrm{lg}(B) \dim(X)$ equations in $\mathrm{Mor}(Y, X)$.

More generally, the same techniques as above yield the following extension of Theorem 2.6 ([Mo1], prop. 2): *locally at a point* $[f]$ *such that* X *is smooth along* $f(Y)$, *the scheme* $\mathrm{Mor}(Y, X; g)$ *can be defined by* $h^1(Y, f^*T_X \otimes I_B)$ *equations in a nonsingular variety of dimension* $h^0(Y, f^*T_X \otimes I_B)$. *In particular, its irreducible components are all of dimension at least*

$$h^0(Y, f^*T_X \otimes I_B) - h^1(Y, f^*T_X \otimes I_B)$$

2.10. Flat families. All this can be done over a Noetherian base scheme S as in [Mo1]: if $Y \to S$ is a *flat* projective S-scheme, $X \to S$ a *flat* quasi-projective S-scheme and B a subscheme of Y *flat* over S with an S-morphism $g : B \to X$, the S-morphisms from Y to X that restrict to g on B can be parametrized by a locally Noetherian S-scheme $\mathrm{Mor}_S(Y, X; g)$. The universal property implies in particular that for any point s of S, one has

$$\mathrm{Mor}_S(Y, X; g)_s \simeq \mathrm{Mor}(Y_s, X_s; g_s)$$

In other words, the schemes $\mathrm{Mor}(Y_s, X_s; g_s)$ fit together to form a scheme over S ([Mo1], prop. 1, and [K1], prop. II.1.5). This is a generalization of the situation we studied in the last paragraph of Section 2.1.

2.11. Morphisms from a curve. Everything takes a particularly simple form when Y is a curve C (and B is finite): for any $f : C \to X$, one has by Riemann–Roch

$$\begin{aligned}
\dim_{[f]} \mathrm{Mor}(C, X) &\geq \chi(C, f^*T_X) \\
&= -K_X \cdot f_*C + (1 - g(C)) \dim(X)
\end{aligned}$$

where $g(C) = 1 - \chi(C, \mathcal{O}_C)$, and

$$\begin{aligned}
\dim_{[f]} \mathrm{Mor}(C, X; f|_B) &\geq \chi(C, f^*T_X) - \mathrm{lg}(B) \dim(X) \qquad (2.2) \\
&= -K_X \cdot f_*C + (1 - g(C) - \mathrm{lg}(B)) \dim(X)
\end{aligned}$$

In the relative situation 2.10, there is also a dimension estimate stated (and proved) in [K1], Theorem II.1.7, for a relative projective flat reduced curve C over an irreducible base S and a smooth quasi-projective S-scheme X. Given a subscheme B of C flat over S, an S-morphism $g : B \to X$ and a morphism $f : C_s \to X_s$ that coincides with g_s on B_s, it says

$$\dim_{[f]} \mathrm{Mor}_S(C, X; g) \geq \chi(C_s, f^*T_{X_s} \otimes I_{B_s}) + \dim(S)$$
$$= -K_{X_s} \cdot f_* C_s + (1 - g(C_s) - \lg(B_s)) \dim(X_s) + \dim(S) \quad (2.3)$$

Furthermore, if $H^1(C_s, f^*T_{X_s} \otimes I_{B_s})$ vanishes, $\mathrm{Mor}_S(C, X; g)$ is smooth over S at $[f]$.[4]

2.12. Morphisms from a curve over a base. Finally, we will need to study deformations of morphisms in the following situation. Let X and Y be projective varieties and let $\pi : X \to Y$ be a morphism. Let C be a curve, let B be a finite subscheme of C, and let $f : C \to X$ be a morphism. We may consider C as a scheme over Y via the morphism $\pi \circ f : C \to Y$, but it is of course not flat in general, so that 2.10 does not apply.

A morphism $f' : C \to X$ is a Y-morphism if and only if $\pi \circ f' = \pi \circ f$. These morphisms are parametrized by the fiber $\mathrm{Mor}_Y(C, X; f|_B)$ of $[\pi \circ f]$ under the map

$$\rho : \mathrm{Mor}(C, X; f|_B) \to \mathrm{Mor}(C, Y; (\pi \circ f)|_B)$$

given by composition by π. If X and Y are *smooth*, it follows from 2.9 that the Zariski tangent space to $\mathrm{Mor}_Y(C, X; f|_B)$ at $[f]$ is the kernel of the tangent map

$$T_{[f]}\rho : H^0(C, f^*T_X \otimes I_B) \to H^0(C, f^*\pi^*T_Y \otimes I_B)$$

Following [Mi], if we let $f^\sharp T_{X/Y}$ be the kernel of

$$f^*T_X \xrightarrow{T\pi} f^*\pi^*T_Y$$

we have

$$T_{\mathrm{Mor}_Y(C, X; f|_B), [f]} \simeq H^0(C, f^\sharp T_{X/Y} \otimes I_B)$$

We will need a lower bound on the dimension of $\mathrm{Mor}_Y(C, X; f|_B)$ analogous to the one obtained in Theorem 2.6. Although this case is not covered by

---

[4]This is not explicitly stated in [K1], but can be deduced from [K1], Theorem II.1.7, as follows. Write h^i for $h^i(C_s, f^*T_{X_s} \otimes I_{B_s})$. If h^1 vanishes, any component of $\mathrm{Mor}_S(C, X; g)$ through $[f]$ has dimension at least $h^0 + \dim(S)$ by [K1], (1.7.2). By [K1], (1.7.1), the Zariski tangent space at $[f]$ to the fiber of $\mathrm{Mor}_S(C, X; g) \to S$ has dimension h^0. It follows that this fiber is smooth at $[f]$ and that every irreducible component of $\mathrm{Mor}_S(C, X; g)$ through $[f]$ has dimension exactly $h^0 + \dim(S)$. By [K1], (1.7.3), $\mathrm{Mor}_S(C, X; g) \to S$ is smooth at $[f]$.

either [Mo1] or [K1] (where it is assumed that B is flat over Y), it is proved in [Mi], th. 1, that *if X, Y and C are smooth, π is surjective and $f(C)$ meets the locus where π is smooth,* we have

$$\dim_{[f]} \mathrm{Mor}_Y(C, X; f|_B) \geq \chi(C, f^\sharp T_{X/Y} \otimes I_B)$$

Let $T_{X/Y}$ be the kernel of the tangent map $T\pi : T_X \to \pi^* T_Y$. Under these hypotheses, there is a morphism $f^* T_{X/Y} \to f^\sharp T_{X/Y}$ which is an isomorphism over the dense open subset U of C inverse image by f of the open set where π is smooth. There is an exact sequence

$$0 \to f^\sharp T_{X/Y} \to f^* T_X \xrightarrow{T\pi} f^* \pi^* T_Y \to Q \to 0$$

where Q is 0 on U, hence is a torsion sheaf on C. It implies

$$
\begin{aligned}
\deg(f^\sharp T_{X/Y}) &= \deg(f^* T_X) + \deg(Q) - \deg(f^* \pi^* T_Y) \\
&\geq \deg(f^* T_X) - \deg(f^* \pi^* T_Y) \\
&= -\deg(f^* K_X) + \deg(f^* \pi^* K_Y) = -\deg(f^* K_{X/Y})
\end{aligned}
$$

The following inequality now follows from Riemann–Roch

$$
\begin{aligned}
\dim_{[f]} \mathrm{Mor}_Y(C, X; f|_B) &\geq \\
&\quad -K_{X/Y} \cdot f_* C + (1 - g(C) - \mathrm{lg}(B))(\dim(X) - \dim(Y)) \quad (2.4)
\end{aligned}
$$

2.4 Lines on a Subvariety of a Projective Space

We will describe lines on complete intersections in a projective space *over an algebraically closed field* **k** to illustrate the concepts developed above.

Let X be a subvariety of $\mathbf{P}^N$. By associating its image to a rational curve, we define a morphism

$$\mathrm{Mor}_1(\mathbf{P}^1, X) \to G(1, \mathbf{P}^N)$$

where $G(1, \mathbf{P}^N)$ is the Grassmannian of lines in $\mathbf{P}^N$. Its image parametrizes lines in X. It has a natural scheme structure and is usually called the Fano[5] variety of X. We will denote it by $F(X)$. It is simpler to study $F(X)$ instead of $\mathrm{Mor}_1(\mathbf{P}^1, X)$.

The induced map $\rho : \mathrm{Mor}_1(\mathbf{P}^1, X) \to F(X)$ is the quotient by the action of the automorphism group of $\mathbf{P}^1$. Let $f : \mathbf{P}^1 \to X$ be a one-to-one parametrization of a line ℓ. *Assume X is smooth of dimension n along ℓ.* Using Proposition 2.4, the tangent map to ρ at $[f]$ fits into an exact sequence

$$0 \to H^0(\mathbf{P}^1, T_{\mathbf{P}^1}) \to H^0(\mathbf{P}^1, f^* T_X) \xrightarrow{T_{[f]}\rho} H^0(\mathbf{P}^1, f^* N_{\ell/X}) \to 0$$

---

[5] This variety is often, but not always, a Fano variety in the sense of Section 3.2.

Since f induces an isomorphism onto its image, we may as well consider the same exact sequence on ℓ. The tangent space to $F(X)$ at $[\ell]$ is therefore $H^0(\ell, N_{\ell/X})$.

Similarly, given a point x on X and a parametrization $f : \mathbf{P}^1 \to X$ of a line contained in X with $f(0) = x$, the group of automorphisms of $\mathbf{P}^1$ fixing 0 acts on the scheme

$$\mathrm{Mor}(\mathbf{P}^1, X; 0 \mapsto x)$$

(notation of 2.9), with quotient the subscheme $F(X, x)$ of $F(X)$ consisting of lines passing through x and contained in X. Lines through x are parametrized by a hyperplane in $\mathbf{P}^N$ of which $F(X, x)$ is a subscheme. From 2.9, it follows that the tangent space to $F(X, x)$ at $[\ell]$ is isomorphic to $H^0(\ell, N_{\ell/X}(-1))$.

There is an exact sequence of normal bundles

$$0 \to N_{\ell/X} \to \mathscr{O}_\ell(1)^{n-1} \to (N_{X/\mathbf{P}^N})|_\ell \to 0 \qquad (2.5)$$

Write

$$N_{\ell/X} \simeq \bigoplus_{i=1}^{n-1} \mathscr{O}_\ell(a_i) \qquad (2.6)$$

where $a_1 \geq \cdots \geq a_{n-1}$. By (2.5), we have $a_1 \leq 1$. If $a_{n-1} \geq -1$, the scheme $F(X)$ is smooth at $[\ell]$ (Theorem 2.6). If $a_{n-1} \geq 0$, the scheme $F(X, x)$ is smooth at $[\ell]$ for any point x on ℓ (see 2.9).

Proposition 2.13 *Let X be a subvariety of $\mathbf{P}^N$ defined by equations of degrees $d_1, \ldots, d_s$. Assume $|\mathbf{d}| = d_1 + \cdots + d_s \leq N - 1$.*

(a) *Through any point of X, there is a line contained in X.*

(b) *If the equations defining X are general and ℓ is a general[6] line contained in X, we have*

$$N_{\ell/X} \simeq \mathscr{O}_\ell(1)^{N-1-|\mathbf{d}|} \oplus \mathscr{O}_\ell^{|\mathbf{d}|-s}$$

PROOF. Assume X is defined by equations $G_1, \ldots, G_s$, with $\deg(G_i) = d_i$. Let x be a point of X with coordinates $(1, 0, \ldots, 0)$. The line joining x to the point $(0, x_1, \ldots, x_N)$ is contained in X if and only if

$$G_i(t, x_1, \ldots, x_N) = 0 \qquad \text{for all } i = 1, \ldots, s$$

Each of these equations is a polynomial in t of degree $d_i - 1$, which has d_i coefficients. Therefore, the set of lines through x contained in X is defined

---

[6] As proved in [K1], th. V.4.3, the scheme $F(X)$ is in this case smooth and irreducible.

by $|\mathbf{d}|$ equations in $\mathbf{P}^{N-1}$. It is nonempty when $|\mathbf{d}| \leq N-1$. This proves item (a).

When X is a complete intersection, its normal bundle is isomorphic to the direct sum $\bigoplus_{i=1}^{s} \mathscr{O}_X(d_i)$. The exact sequence (2.5) yields

$$0 \to N_{\ell/X}(-1) \to \mathscr{O}_{\ell}^{N-1} \to \bigoplus_{i=1}^{s} \mathscr{O}_{\ell}(d_i - 1) \to 0$$

Assume that the line ℓ has equations $x_2 = \cdots = x_N = 0$. The map

$$\alpha : H^0(\ell, \mathscr{O}_{\ell})^{N-1} \to \bigoplus_{i=1}^{s} H^0(\ell, \mathscr{O}_{\ell}(d_i - 1))$$

is given by

$$\alpha(\lambda_2, \ldots, \lambda_n) = \left(\sum_{j=2}^{N} \lambda_j \left(\frac{\partial G_1}{\partial x_j} \right)_{|\ell}, \ldots, \sum_{j=2}^{N} \lambda_j \left(\frac{\partial G_s}{\partial x_j} \right)_{|\ell} \right)$$

Write $G_i = \sum_{j=2}^{N} x_j G_{ij}$. The matrix of α is

$$\begin{pmatrix} (G_{12}) & \cdots & (G_{1N}) \\ (G_{22}) & \cdots & (G_{2N}) \\ \vdots & & \vdots \\ (G_{s2}) & \cdots & (G_{sN}) \end{pmatrix}$$

where (G_{ij}) is the column of the coefficients of the homogeneous polynomial G_{ij} in some basis of $H^0(\ell, \mathscr{O}_{\ell}(d_i - 1))$. It can therefore be any matrix hence will have maximal rank for general $G_1, \ldots, G_s$. This proves (b). $\quad\square$

In case (b) of the proposition, Theorem 2.6 implies that the scheme $F(X)$ is in particular smooth at $[\ell]$ of dimension

$$|\mathbf{d}| - s + 2(N - 1 - |\mathbf{d}|) = 2N - 2 - s - |\mathbf{d}|$$

We will now see on a very simple example that this does not always hold, even when X is smooth (but not general).

2.14. Fermat hypersurfaces. Let X be a smooth hypersurface in $\mathbf{P}^N$ defined by a polynomial G of degree d. Assume X contains a line ℓ. It follows from the proof above that $H^1(\ell, N_{\ell/X}(-1))$ is the cokernel of the map

$$\begin{aligned} \alpha' : \quad & \mathbf{k}^{N+1} & \to \quad & H^0(\ell, \mathscr{O}_{\ell}(d-1)) \\ & (\lambda_0, \ldots, \lambda_N) & \mapsto \quad & \sum_{j=0}^{N} \lambda_j \left(\frac{\partial G}{\partial x_j} \right)_{|\ell} \end{aligned}$$

The Fermat hypersurface X_N^d is the hypersurface in $\mathbf{P}^N$ defined by the equation

$$x_0^d + \cdots + x_N^d = 0$$

It is smooth if and only if the characteristic of $\mathbf{k}$ does not divide d, a condition that we will always assume to hold.

There are "obvious" lines contained in X_N^d. They can be described as follows. Fix a partition of $\{0, \ldots, N\}$ into subsets $I_1, \ldots, I_r$, each with at least two elements. To each point $x = (x_{I_1}, \ldots, x_{I_r})$ in $\mathbf{P}^N$ such that

$$\sum_{i \in I_j} x_i^d = 0 \qquad \text{for all } j$$

corresponds a linear subspace

$$\{(\lambda_1 x_{I_1}, \ldots, \lambda_r x_{I_r}) \mid (\lambda_1, \ldots, \lambda_r) \in \mathbf{P}^{r-1}\}$$

of dimension $r - 1$ contained in X_N^d. We obtain a family of linear spaces parametrized by $\prod_{j=1}^r X_{\mathrm{Card}(I_j)-1}^d$, which has dimension

$$\sum_{j=1}^r (\mathrm{Card}(I_j) - 2) = N + 1 - 2r$$

Each such linear space in turn contains a $(2r - 4)$-dimensional family of lines. Altogether, we get a $(N - 3)$-dimensional family of lines contained in X_N^d. We will call these lines *standard*. We will show in Exercise 2.5.3 that when $d \geq N$ and the characteristic of $\mathbf{k}$ is either 0 or $> d$, all lines contained in X_N^d are standard.

Consider a standard line ℓ constructed from a partition of $\{0, \ldots, N\}$ into two subsets. The map α' above has rank 2: its image is generated by the polynomials t^{d-1} and u^{d-1}. It follows, using Riemann–Roch,

$$h^1(\ell, N_{\ell/X_N^d}(-1)) = d - 2 \qquad h^0(\ell, N_{\ell/X_N^d}(-1)) = N - 3$$

With the notation of (2.6), this implies that the number of a_i equal to 1 is $N - 2$, hence

$$N_{\ell/X_N^d} \simeq \mathcal{O}_\ell(1)^{N-3} \oplus \mathcal{O}_\ell(2 - d) \qquad (2.7)$$

(see Exercise 2.5.3 for more).

2.15. *Assume the characteristic p of $\mathbf{k}$ is positive and $d = p^r + 1$. The line joining two points x and y is contained in X_N^d if and only if*

$$
\begin{aligned}
0 &= \sum_{j=0}^{N} (x_j + ty_j)^{p^r+1} \\
&= \sum_{j=0}^{N} (x_j^{p^r} + t^{p^r} y_j^{p^r})(x_j + ty_j) \\
&= \sum_{j=0}^{N} (x_j^{p^r+1} + tx_j^{p^r} y_j + t^{p^r} x_j y_j^{p^r} + t^{p^r+1} y_j^{p^r+1})
\end{aligned}
\tag{2.8}
$$

In particular, there is an $(N-4)$-dimensional family of lines through x contained in X_N^d, which is therefore covered by lines if $N \geq 4$.

This calculation also implies that $F(X_N^d)$ has dimension at least $2N - 6$ at every point. By Theorem 2.6, the normal bundle to *any* line contained in X_N^d must have at least $2N - 6$ independent sections, hence decomposition type (2.7), which is *not* the general decomposition type of Proposition 2.13(b). By Proposition 2.6, the tangent space to $F(X_N^d)$ has dimension $2N - 6$, hence $F(X_N^d)$ *is smooth of dimension* $2N - 6$ (it is also connected when $N \geq 4$ by Exercise 2.5.2).[7]

Also, given any line ℓ contained in X_N^d and passing through x, we have, with the notation introduced before Proposition 2.13,

$$
\dim_{[\ell]} F(X_N^d, x) \geq N - 4
$$

and there is equality for x general, because

$$
\dim(F(X_N^d, x)) = \dim(F(X_N^d)) + 2 - N
$$

On the other hand (see 2.9)

$$
T_{F(X_N^d, x), [\ell]} \simeq H^0(\ell, N_{\ell/X_N^d}(-1)) \simeq \mathbf{k}^{N-3}
$$

so that $F(X_N^d, x)$ *is nowhere reduced*. This can also be seen directly: by (2.8), this scheme is defined by the equations

$$
0 = \sum_{j=0}^{N} x_j^{p^r} y_j = \left(\sum_{j=0}^{N} x_j^{p^{-r}} y_j \right)^{p^r} = \sum_{j=0}^{N} y_j^{p^r+1}
$$

---

[7]This phenomenon does not happen in characteristic zero, at least when N is (very very) large with respect to d: it was shown in [HMP] and [Ch] that in that case, Proposition 2.13(b) holds whenever X is smooth (and ℓ is general). In the same direction, J. de Jong and I have conjectured that $F(X)$ should have the expected dimension $2N - 3 - d$ whenever X is a smooth hypersurface of degree $d \leq N$ and the characteristic of $\mathbf{k}$ is $\geq d$.

where y is in some fixed hyperplane not containing x.

2.16. *Assume $N = 3$ and $d \geq 4$.* A direct calculation (see [AK], prop 1.3 or Exercise 2.5.3) shows that if the characteristic of **k** is either 0 or at least $d + 1$, all lines contained in X_3^d are of the type described above in 2.14 (or are obtained from those by permutating the coordinates), so that $F(X_3^d)$ has $10d$ irreducible components, all of dimension 1. However,

$$T_{F(X_3^d), [\ell]} \simeq H^0(\ell, N_{\ell/X_3^d}) \simeq \mathbf{k}^2$$

so that $F(X_3^d)$ *is nowhere reduced* (see Exercise 2.5.3 for more details).

2.5 Exercises

All varieties are defined over an algebraically closed field **k** of characteristic p. Recall (see 2.14) that the Fermat hypersurface X_N^d has equation

$$x_0^d + \cdots + x_N^d = 0$$

in $\mathbf{P}^N$ (we always assume that p does not divide d).

The purpose of this series of exercises is to show (among other things) the following.

(a) *Assume $p = 0$ or $p > d$.*

- *For $d < N$, the scheme $F(X_N^d)$ of lines in X_N^d has the expected dimension $2N - 3 - d$. It is irreducible and reduced.*

- *When $d \geq N$, the scheme $F(X_N^d)$ has dimension $N - 3$ and all lines on X_N^d can be explicitly described. When $N \geq 4$, the scheme $F(X_N^d)$ is reducible and nonreduced.*[8]

(b) *Assume $d = p^r + 1$ and $N \geq 3$. The variety X_N^d is unirational and $F(X_N^d)$ is smooth of dimension $2N - 6$, connected when $N \geq 4$.*

1. Assume $p > 0$ and $N \geq 3$. For every dth root ω of -1, the hypersurface X_N^d contains the line ℓ joining the points $(1, \omega, 0, 0, \ldots, 0)$ and $(0, 0, 1, \omega, 0, \ldots, 0)$. The pencil

$$-t\omega x_0 + t x_1 - \omega x_2 + x_3 = 0$$

of hyperplanes containing ℓ induces a rational map $\pi : X_N^d \dashrightarrow \mathbf{A}_{\mathbf{k}}^1$ which makes $k(X_N^d)$ an extension of $\mathbf{k}(t)$.

---

[8]More precisely, each component is nonreduced when N is even and there are generically reduced and generically nonreduced components when N is odd.

(a) Assume that $d-1$ is a power q of p. Show that the generic fiber of π is isomorphic over $k(t^{1/q})$ to

- if $N = 3$, the rational plane curve with equation

$$y_2^{q-1} y_3 + y_1^q = 0$$

- if $N \geq 4$, the singular rational hypersurface with equation[9]

$$y_2^q y_3 + y_2 y_1^q + y_4^{q+1} + \cdots + y_N^{q+1} = 0$$

in $\mathbf{P}^{N-1}$.

Deduce that X_N^{q+1} has a purely inseparable cover of degree q which is rational.

(b) Show that X_N^d is unirational whenever $N \geq 3$ and d divides $p^r + 1$ for some positive integer r.

2. Assume $p > 0$. The scheme $F(X_N^{p^r+1})$ was shown to be smooth of dimension $2N - 6$ in 2.15. Show that it is connected for $N \geq 4$. (*Hint:* For $N \geq 5$ and x general, show that the scheme $F(X_N^{p^r+1}, x)_{\mathrm{red}}$ is smooth and connected (see [B2]).)

3. Throughout this exercise, we assume $p = 0$ or $p > d$.

(a) Assume $d \geq 3$. Show that all lines on X_3^d are standard (see 2.14). There are therefore $3d^2$ of them. However, if p is positive and $d - 1$ is a power of p, show that there are exactly $d^3(d-3)$ nonstandard lines on X_3^d.

(b) Let ℓ be a standard line contained in X_N^d, which is general in its $\mathbf{P}^{r-1}$. Show

$$h^0(\ell, N_{\ell/X_N^d}(-1)) = N - r - 1 \qquad h^0(\ell, N_{\ell/X_N^d}) = 2(N - r - 1)$$

When $r = 2$, we have from (2.7)

$$N_{\ell/X_N^d} \simeq \mathcal{O}_\ell(1)^{N-3} \oplus \mathcal{O}_\ell(2 - d)$$

When $r = 3$ (hence $N \geq 5$), show

$$N_{\ell/X_N^d} \simeq \mathcal{O}_\ell(1)^{N-4} \oplus \mathcal{O}_\ell\left(1 - \frac{d}{2}\right) \oplus \mathcal{O}_\ell\left(2 - \frac{d}{2}\right)$$

when d is even and

$$N_{\ell/X_N^d} \simeq \mathcal{O}_\ell(1)^{N-4} \oplus \mathcal{O}_\ell\left(1 - \frac{d-1}{2}\right) \oplus \mathcal{O}_\ell\left(1 - \frac{d-1}{2}\right)$$

when d is odd.

---

[9]This shows in particular that the "Remarque" on p. 125 of [B2] is incorrect!

(c) Assume $d \geq N$. Show that all lines on X_N^d are standard. (*Hint:* Proceed by induction on N and show first that any line contained in X_N^d has a point with at least two coordinates equal to 0. Use this point to construct a line contained in X_{N-1}^d.)

It follows that all components of $F(X_N^d)$ have dimension $N - 3$, that their number is the number of partitions of $\{0, \ldots, N\}$ into subsets with at least two elements, where each partition is counted d^s times, s being the number of subsets with two elements (this is because X_1^d is a set with d elements).

Show that the only generically reduced components correspond to the case $N + 1 = 2r = 2s$. (*Hint:* Use (b).)

(d) For $d = N$, it follows from (c) that the scheme $F(X_N^d)$ has expected dimension $N - 3$. Its class in the Grassmannian $G(1, \mathbf{P}^N)$ was calculated in [DM]. Here are a few values:[10]

$$\begin{aligned}
[F(X_4^4)] &= 320\sigma_{3,2} \\
[F(X_5^5)] &= 25(130\sigma_{4,2} + 115\sigma_{3,3}) \\
[F(X_6^6)] &= 36(924\sigma_{5,2} + 3360\sigma_{4,3})
\end{aligned}$$

Show that $F(X_4^4)$ is the union of 40 isomorphic curves that each have multiplicity 2.

Show that $F(X_5^5)$ is the union of

- 75 isomorphic surfaces with multiplicity 6 and class $5\sigma_{4,2}$;
- 10 isomorphic surfaces with multiplicity 4 and class $25(\sigma_{4,2} + \sigma_{3,3})$;
- 1875 isomorphic reduced surfaces with class $\sigma_{3,3}$.

Show that $F(X_6^6)$ is the union of

- 126 isomorphic threefolds with class $6\sigma_{5,2}$;
- 35 isomorphic threefolds with class $36(\sigma_{5,2} + \sigma_{4,3})$;
- 3780 isomorphic threefolds with class $6\sigma_{4,3}$;

and that the respective multiplicities are either 4, 24, 4 or 34, 6, 5.

---

[10]There is an error in the table in [DM]: a 91 needs to be changed into 115.

3

"Bend-and-Break" Lemmas

We now enter Mori's world. The whole story began in 1979, with Mori's spectacular proof of a conjecture of Hartshorne characterizing projective spaces as the only smooth projective varieties with ample tangent bundle. The techniques that Mori introduced to solve this conjecture have turned out to have more far-reaching applications than Hartshorne's conjecture itself.

Mori's first idea is that if a curve deforms on a projective variety X while passing through a fixed point, it must at some point break up with at least one rational component, hence the name "bend-and-break." This is a relatively easy result, but now comes the really tricky part: when X is smooth, to ensure that a morphism $f : C \to X$ deforms fixing a point, the natural thing to do is to use the lower bound (2.2)

$$-K_X \cdot f_*C - g(C)\dim(X)$$

for the dimension of the space of deformations. How can one make this number positive? The divisor $-K_X$ had better have some positivity property, but even if it does, simple-minded constructions like ramified covers never lead to a positive bound. Only in positive characteristic can Frobenius operate its magic: increase the degree of f (hence the intersection number $-K_X \cdot f_*C$ if it is positive) *without changing the genus of C.*

The most favorable situation is when X is a Fano variety, which means that $-K_X$ is ample: in that case, any curve has positive $(-K_X)$-degree and the Frobenius trick combined with Mori's bend-and-break lemma produces a rational curve through any point of X. Another bend-and-break-type result universally bounds the $(-K_X)$-degree of this rational curve and allows a proof in all characteristics of the fact that Fano varieties are covered by rational curves by reducing to the positive characteristic case (Theorem 3.4).

We then prove a finer version of the bend-and-break lemma (Proposition 3.5) and deduce a result that will be essential for the description of the cone of curves of any projective smooth variety (Theorem 6.1): if K_X has negative degree on a curve C, the variety X contains a rational curve that meets C (Theorem 3.6). We give a direct application in Theorem 3.10 by showing that varieties for which $-K_X$ is nef but not numerically trivial are also covered by rational curves.

The bend-and-break lemma also has relative versions. Assume our variety X comes equipped with a morphism $\pi : X \to Y$. Two things can happen to the deformations of the curve C fixing a point: either the images in Y do not vary, or they do. In the first case, we produce a rational curve in X contracted by π (Proposition 3.11), and in the second case, a rational curve in X *not* contracted by π (Proposition 3.20).

We draw (reducing again to positive characteristic and using the Frobenius trick) various consequences from these two results. The first result implies that in some cases, $-K_Y$ inherits various positivity properties of $-K_X$ (Corollaries 3.14 and 3.15). More importantly, it also implies that the Albanese map of a smooth variety with nef anticanonical divisor is surjective and has connected fibers, a result of Zhang proved in Theorem 3.17.

The second result is used to prove the existence, on a Fano variety X, of rational curves that are transverse to any given nonconstant morphism $X \to Y$ (Corollaries 3.21 and 3.22). This result will be one of the two main ingredients of our proof that any two points of a Fano variety can be joined by a chain of rational curves (Proposition 5.16).

We work here over an *algebraically closed field* **k**.

Recall (§1.1) that a 1-cycle on X is a formal sum $\sum_{i=1}^{s} n_i C_i$, where the n_i are integers and the C_i are integral curves on X. It is called rational if the C_i are rational curves. If C is a curve with irreducible components $C_1, \ldots, C_r$ and $f : C \to X$ a morphism, we will write $f_* C$ for the effective 1-cycle $\sum_{i=1}^{r} (\deg f|_{C_i}) f(C_i)$.

3.1   Producing Rational Curves

The following is the original bend-and-break lemma, which can be found in [Mo1] (th. 5 and 6). It says that a curve deforming nontrivially, while keeping a point fixed, must break into an effective 1-cycle with a rational component passing through the fixed point.

Proposition 3.1 *Let X be a projective variety, let $f : C \to X$ be a smooth curve, and let c be a point on C. If $\dim_{[f]} \mathrm{Mor}(C, X; f|_{\{c\}}) \geq 1$, there exists a rational curve on X through $f(c)$.*

According to (2.2), *when X is smooth along $f(C)$*, the hypothesis is fulfilled whenever

$$-K_X \cdot f_* C - g(C)\dim(X) \geq 1$$

The proof actually shows that there exists a morphism $f' : C \to X$ and a connected nonzero effective rational 1-cycle Z on X passing through $f(c)$ such that

$$f_* C \sim f'_* C + Z$$

PROOF OF THE PROPOSITION. We may assume that C is irrational. Let T be the normalization of a 1-dimensional subvariety of $\operatorname{Mor}(C, X; f|_{\{c\}})$ passing through $[f]$ and let $\overline{T}$ be a smooth compactification of T. The indeterminacies of the rational map

$$\mathrm{ev} : C \times \overline{T} \dashrightarrow X$$

coming from the morphism $T \to \operatorname{Mor}(C, X; f|_{\{c\}})$ can be resolved by blowing up points to get a morphism

$$e : S \xrightarrow{\varepsilon} C \times \overline{T} \xrightarrow{\mathrm{ev}} X$$

The following picture sums up our constructions.

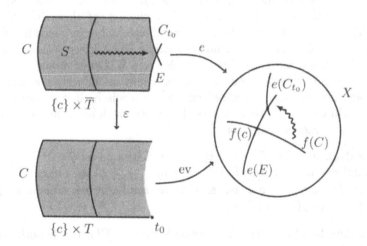

The curve $f(C)$ degenerates to a 1-cycle with a rational component $e(E)$.

If ev is defined at every point of $\{c\} \times \overline{T}$, take an affine open neighborhood U of $f(c)$ in X and an open neighborhood V of c in C such that $\mathrm{ev}(V \times \overline{T})$ is contained in U. For each v in V, the image of $\{v\} \times \overline{T}$ is a complete subvariety of the affine variety U, hence is a point. It follows that ε has infinitely many 1-dimensional fibers, so that its image must be the curve $f(C)$. In particular, for each t in T, the image of f_t must be $f(C)$. But C being irrational, the pair (C, c) has only finitely many automorphisms, so this would imply that the f_t remain the same, which is absurd.

Hence there exists a point t_0 in $\overline{T}$ such that ev is not defined at (c, t_0). The fiber of t_0 under the projection $S \to \overline{T}$ is the union of the strict transform of $C \times \{t_0\}$ and a (connected) exceptional rational 1-cycle E which is not entirely contracted by e and meets the strict transform of $\{c\} \times \overline{T}$. Since the latter is contracted by e to the point $f(c)$, the rational 1-cycle $e_* E$ passes through $f(c)$. $\square$

The same proof shows that the proposition still holds when X is a complex compact Kähler variety (because, by a result of Campana, the variety $\mathrm{Mor}(C, X; f|_{\{c\}})$ is *Moishezon*, hence contains curves as soon as its dimension is positive). However, it fails in general for curves on compact complex manifolds. An example can be constructed as follows: let E be an elliptic curve, let $\mathscr{L}$ be a very ample line bundle on E, and let s and s' be sections of $\mathscr{L}$ that generate it at each point. The sections (s, s'), $(is, -is')$, $(s', -s)$, and (is', is) of $\mathscr{L} \oplus \mathscr{L}$ are independent over $\mathbf{R}$ in each fiber. They generate a discrete subgroup of the total space of $\mathscr{L} \oplus \mathscr{L}$ and the quotient X is a compact complex manifold with a morphism $\pi : X \to E$ whose fibers are 2-dimensional complex tori. There is a 1-dimensional family of sections $\sigma_t : E \to X$ of π defined by $\sigma_t(x) = (ts(x), 0)$, for $t \in \mathbf{C}$, and they all pass through the points of the zero section where s vanishes. However, X contains no rational curves, because it would have to be contained in a fiber of π, and complex tori contain no such curve. The variety X is of course not algebraic, and not even bimeromorphic to a Kähler manifold.

Once we know there is a rational curve, it may under certain conditions be broken up into several components. More precisely, if it deforms nontrivially while keeping two points fixed, it must break up (into an effective 1-cycle with rational components).

Proposition 3.2 *Let X be a projective variety and let $f : \mathbf{P}^1 \to X$ be a rational curve. If $\dim_{[f]}(\mathrm{Mor}(\mathbf{P}^1, X; f|_{\{0, \infty\}})) \geq 2$, the 1-cycle $f_* \mathbf{P}^1$ is numerically equivalent to a connected nonintegral effective rational 1-cycle passing through $f(0)$ and $f(\infty)$.*

According to (2.2), *when X is smooth along $f(\mathbf{P}^1)$, the hypothesis is fulfilled whenever*

$$-K_X \cdot f_* \mathbf{P}^1 - \dim(X) \geq 2$$

PROOF OF THE PROPOSITION. The group of automorphisms of $\mathbf{P}^1$ fixing two points is the multiplicative group $\mathbf{G}_m$. Let T be the normalization of a 1-dimensional subvariety of $\mathrm{Mor}(\mathbf{P}^1, X; f|_{\{0, \infty\}})$ passing through $[f]$ but not contained in its $\mathbf{G}_m$-orbit. The corresponding map

$$F : \mathbf{P}^1 \times T \to X \times T$$

is finite. Let $\overline{T}$ be a smooth compactification of T and let S be the normalization in $K(\mathbf{P}^1 \times T)$ of the closure in $X \times \overline{T}$ of the image of F, with $\overline{F} : S \to X \times \overline{T}$ the canonical finite morphism.[1] Since $\mathbf{P}^1 \times T$ is already normal, we have $\overline{F}^{-1}(X \times T) = \mathbf{P}^1 \times T$ by uniqueness of normalizations, so that there is a commutative diagram

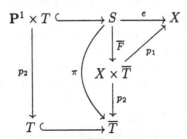

This construction is similar to the one we performed in the last proof and can be summed up in the following picture.

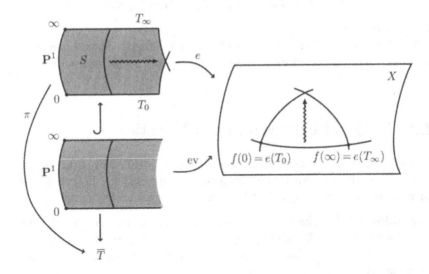

The rational 1-cycle $f(C)$ bends and breaks.

The surface S might not be smooth. On the other hand, we know that no component of a fiber of π is contracted by e (because it would then be contracted by $\overline{F}$).

Since $\overline{T}$ is a smooth curve and S is integral, π is flat ([H1], III, prop. 9.7); hence each fiber C is a 1-dimensional projective scheme without embedded

---

[1]This is constructed exactly as the standard normalization (see [H1], II, ex. 3.8) by patching up the spectra of the integral closures *in* $K(\mathbf{P}^1 \times T)$ of the coordinate rings of affine open subsets of $\overline{F(\mathbf{P}^1 \times T)}$. The fact that $\overline{F}$ is finite follows from the finiteness of integral closure ([H1], I, th. 3.9A).

component, whose genus is constant hence equal to 0 ([H1], III, cor. 9.10). In particular, any component C_1 of C_{red} is smooth rational, because $\mathcal{O}_{C_1}$ is a quotient of $\mathcal{O}_C$; hence $H^1(C_1, \mathcal{O}_{C_1})$ is a quotient of $H^1(C, \mathcal{O}_C)$, and therefore vanishes. In particular, if C is integral, it is a smooth rational curve.

Assume all fibers of π are integral. Then S is a (minimal) ruled surface in the sense of [H1], V, §2 (Hartshorne assumes that S is smooth, but this hypothesis is not used in the proofs, hence follows from the others). Let T_0 be the closure of $\{0\} \times T$ in S and let T_∞ be the closure of $\{\infty\} \times T$. These sections of π are contracted by e (to $f(0)$ and $f(\infty)$, respectively).

If H is an ample divisor on $e(S)$, which is a surface by construction, we have $(e^*H)^2 > 0$ and $e^*H \cdot T_0 = e^*H \cdot T_\infty = 0$; hence T_0^2 and T_∞^2 are negative by the Hodge index theorem.

However, since T_0 and T_∞ are both sections of π, their difference is linearly equivalent to the pull-back by π of a divisor on $\overline{T}$ ([H1], V, prop. 2.3). In particular,

$$0 = (T_0 - T_\infty)^2 = T_0^2 + T_\infty^2 - 2T_0 \cdot T_\infty < 0$$

which is absurd.

It follows that at least one fiber of π is not integral. Since none of its components is contracted by e, its direct image on X is the required 1-cycle. $\square$

3.2 Rational Curves on Fano Varieties

A Fano variety is a smooth projective variety X (defined over the algebraically closed field $\mathbf{k}$) with ample anticanonical divisor; K_X is therefore as far as possible from being nef: it has negative degree on any curve.

Examples 3.3 (1) The projective space is a Fano variety. A smooth complete intersection in $\mathbf{P}^n$ defined by equations of degrees $d_1, \ldots, d_s$ with $d_1 + \cdots + d_s \leq n$ is a Fano variety. A finite product of Fano varieties is a Fano variety.

(2) If Y is a Fano variety and $D_1, \ldots, D_r$ are nef divisors on Y such that $-K_Y - D_1 - \cdots - D_r$ is ample, $X = \mathbf{P}(\bigoplus_{i=1}^r \mathcal{O}_Y(D_i))$ is a Fano variety.[2] Indeed, if D is a divisor associated with the line bundle $\mathcal{O}_Y(1)$ and π is the canonical map $X \to Y$, one gets as in [H1] (V, Lemma 2.10),

$$-K_X = (r+1)D + \pi^*(-K_Y - D_1 - \cdots - D_r)$$

Since each D_i is nef, the divisor D is nef on X. Since each $-K_Y - D_1 - \cdots - D_r + D_i$ is ample (1.22), the divisor $D + \pi^*(-K_Y - D_1 - \cdots - D_r)$

---

[2] We follow Grothendieck's notation: for a vector bundle $\mathcal{E}$, the projectivization $\mathbf{P}(\mathcal{E})$ is the space of *hyperplanes* in the fibers of $\mathcal{E}$.

is ample (because a direct sum of nef (resp. ample) line bundles is a vector bundle which has the same property). It follows that $-K_Y$ is ample (1.22).

For instance, the variety $Y_{r,s}$ constructed in 1.36 is a Fano variety when $0 \leq r < s$.

We will apply the bend-and-break lemmas to show that any Fano variety X is covered by rational curves. We start from any curve $f : C \to X$ and want to show, using the estimate (2.2), that it deforms nontrivially while keeping a point x fixed. As explained in the introduction, we only know how to do that in positive characteristic, where the Frobenius morphism allows us to increase the degree of f without changing the genus of C. This gives in that case the required rational curve through x. Using the second bend-and-break lemma, we can bound the degree of this curve by a constant depending only on the dimension of X, and this will be essential for the remaining step: reduction of the characteristic zero case to positive characteristic.

Assume for a moment that X and x are defined over $\mathbf{Z}$. For almost all prime numbers p, the reduction of X modulo p is a Fano variety of the same dimension; hence there is a rational curve (defined over the algebraic closure of $\mathbf{F}_p$) through x. This means that the scheme $\mathrm{Mor}_{>0}(\mathbf{P}^1, X; 0 \to x)$, which is defined over $\mathbf{Z}$, has a geometric point modulo almost all primes p. Since we can moreover bound the degree of the curve by a constant independent of p, we are in fact dealing with a quasi-projective scheme, and this implies that it has a point over $\bar{\mathbf{Q}}$, hence over $\mathbf{k}$ (this is essentially because a system of polynomial equations with integral coefficients that has a solution modulo almost all primes has a solution). In general, X and x are defined over some finitely generated ring and a similar reasoning yields the existence of a k-point of $\mathrm{Mor}_{>0}(\mathbf{P}^1, X; 0 \to x)$, that is, of a rational curve on X through x.

Theorem 3.4 *Let X be a Fano variety of positive dimension n. Through any point of X there is a rational curve of $(-K_X)$-degree at most $n + 1$.*

There is no known proof of this result (say over the complex numbers) that uses only transcendental methods.

PROOF OF THE THEOREM. Let x be a point of X. To construct a rational curve through x, it is enough by Proposition 3.1 to produce a curve $f : C \to X$ and a point c on C with $\dim_{[f]} \mathrm{Mor}(C, X; f|_{\{c\}}) \geq 1$ and $f(c) = x$. By the dimension estimate of (2.2), it is enough to have

$$-K_X \cdot f_*C - ng(C) \geq 1$$

Unfortunately, there is no known way to achieve that, except in positive characteristic. Here is how it works.

Assume that the field $\mathbf{k}$ has characteristic $p > 0$. Choose a smooth curve $f : C \to X$ through x and a point c of C such that $f(c) = x$. Consider the (k-linear) Frobenius morphism $C_1 \to C$ ([H1], pp. 301–302). It has degree p, but C_1 and C being isomorphic as abstract schemes have the same genus. Iterating the construction, we get a morphism $F_m : C_m \to C$ of degree p^m between curves of the same genus. But

$$-K_X \cdot (f \circ F_m)_* C_m - ng(C_m) = -p^m K_X \cdot f_* C - ng(C)$$

is positive for m large enough. By Proposition 3.1, there exists a rational curve $f' : \mathbf{P}^1 \to X$, with say $f'(0) = x$. If

$$-K_X \cdot f'_* \mathbf{P}^1 - n \geq 2$$

the scheme $\mathrm{Mor}(\mathbf{P}^1, X; f'|_{\{0,1\}}))$ has dimension at least 2 at $[f']$. By Proposition 3.2, one can break up the rational curve $f'(\mathbf{P}^1)$ into at least two (rational) pieces. The component passing through x has smaller $(-K_X)$-degree, and we can repeat the process as long as $-K_X \cdot \mathbf{P}^1 - n \geq 2$, until we get a rational curve of degree no more than $n + 1$.

This proves the theorem in positive characteristic. Assume now that $\mathbf{k}$ has characteristic 0. Embed X in some projective space, where it is defined by a finite set of equations, and let R be the (finitely generated) subring of $\mathbf{k}$ generated by the coefficients of these equations and the coordinates of x. There is a projective scheme $\mathscr{X} \to \mathrm{Spec}(R)$ with an R-point x_R, such that X is obtained from its generic fiber by base change from the quotient field of R to $\mathbf{k}$.

The geometric generic fiber is a Fano variety of dimension n. There is a dense open subset U of $\mathrm{Spec}(R)$ over which $\mathscr{X}$ is flat ([Gr3], th. 6.9.1), and even smooth of dimension n ([Gr4], th. 12.2.4(iii)). Since ampleness is an open property ([Gr4], cor. 9.6.4), we may even, upon shrinking U, assume that the relative dualizing sheaf $\omega_{\mathscr{X}_U/U}$ is ample on each fiber. It follows that for each maximal ideal $\mathfrak{m}$ of R in U, the (geometric) fiber $X_\mathfrak{m}$ is a Fano variety of dimension n. In the future, we will skip these steps when we use this process of reduction to positive characteristic.

Let us take a short break and use a little commutative algebra to show that the finitely generated ring R has the following properties:

- for each maximal ideal $\mathfrak{m}$ of R, the field $R/\mathfrak{m}$ is finite;

- maximal ideals are dense in $\mathrm{Spec}(R)$.

The first item is proved as follows. The field $R/\mathfrak{m}$ is a finitely generated $(\mathbf{Z}/\mathbf{Z} \cap \mathfrak{m})$-algebra, hence is finite over the quotient field of $\mathbf{Z}/\mathbf{Z} \cap \mathfrak{m}$ by a theorem of Zariski.[3] If $\mathbf{Z} \cap \mathfrak{m} = 0$, the field $R/\mathfrak{m}$ is a finite-dimensional $\mathbf{Q}$-vector space with basis $e_1, \ldots, e_m$. If $x_1, \ldots, x_r$ generate the $\mathbf{Z}$-algebra

---

[3]This theorem says that if k is a field and K a finitely generated k-algebra which is a field, K is an algebraic hence finite extension of k; see [Ma], th. 5.2.

$R/\mathfrak{m}$, there exists an integer q such that qx_j belongs to $\mathbf{Z}e_1 \oplus \cdots \oplus \mathbf{Z}e_m$ for each j. This implies

$$\mathbf{Q}e_1 \oplus \cdots \oplus \mathbf{Q}e_m = R/\mathfrak{m} \subset \mathbf{Z}[1/q]e_1 \oplus \cdots \oplus \mathbf{Z}[1/q]e_m$$

which is absurd. Therefore, $\mathbf{Z}/\mathbf{Z} \cap \mathfrak{m}$ is finite and so is $R/\mathfrak{m}$.

For the second item, we need to show that the intersection of all maximal ideals of R is $\{0\}$. Let a be a nonzero element of R and let $\mathfrak{n}$ be a maximal ideal of the localization R_a. The field $R_a/\mathfrak{n}$ is finite by the first item; hence its subring $R/R \cap \mathfrak{n}$ is a finite domain, hence a field. Therefore, $R \cap \mathfrak{n}$ is a maximal ideal of R, which is in the open subset $\mathrm{Spec}(R_a)$ of $\mathrm{Spec}(R)$.

Now back to the proof of the theorem. As proved in 1.10, there is a quasi-projective scheme

$$\rho : \bigsqcup_{0 < d \leq n+1} \mathrm{Mor}_d(\mathbf{P}_R^1, \mathscr{X}; 0 \mapsto x_R) \to \mathrm{Spec}(R)$$

that parametrizes nonconstant morphisms of degree at most $n + 1$.

Let $\mathfrak{m}$ be a maximal ideal of R. Since the field $R/\mathfrak{m}$ is finite, hence of positive characteristic, what we just saw implies that the (geometric) fiber over a closed point of the dense open subset U of $\mathrm{Spec}(R)$ is nonempty. It follows that the image of ρ, which is a constructible[4] subset of $\mathrm{Spec}(R)$ by Chevalley's theorem ([H1], II, ex. 3.19), contains all closed points of U. It is therefore dense by the second item, hence contains the generic point ([H1], II, ex. 3.18(b)). This implies that the generic fiber is nonempty. It has therefore a geometric point, which corresponds to a rational curve on X through x, of degree at most $n + 1$, defined over an algebraic closure of the quotient field of R, hence over $\mathbf{k}$.[5]   □

3.3   A Stronger Bend-and-Break Lemma

We prove a generalization of the bend-and-break lemma (Proposition 3.1) that gives some control over the degree of the rational curve that is produced.

Starting from a curve that deforms nontrivially with a finite (nonzero) number of fixed points, we produce a rational curve that passes through at least one of the fixed points and get an upper bound on its degree with respect to a fixed ample divisor. The more points are fixed, the better the bound on the degree is. The ideas are the same as in the original bend-and-break, with additional computations of intersection numbers thrown in.

---

[4]Recall that a constructible subset is a finite union of locally closed subsets.

[5]It is important to remark that the "universal" bound on the degree of the rational curve is essential for the proof.

This result first appeared in [MM].

Proposition 3.5 *Let X be a projective variety and let H be an ample divisor on X. Let $f : C \to X$ be a smooth curve and let B be a finite nonempty subset of C. Assume*

$$\dim_{[f]} \operatorname{Mor}(C, X; f|_B) \geq 1$$

There exists a rational curve Γ on X that meets $f(B)$ and such that

$$H \cdot \Gamma \leq \frac{2H \cdot C}{\operatorname{Card}(B)}$$

According to (2.2), *when X is smooth along $f(C)$,* the hypothesis is fulfilled whenever

$$-K_X \cdot f_*C + (1 - g(C) - \operatorname{Card}(B)) \dim(X) \geq 1$$

The proof actually shows that there exist a morphism $f' : C \to X$ and a nonzero effective rational 1-cycle Z on X such that

$$f_*C \sim f'_*C + Z$$

one component of which meets $f(B)$ and satisfies the degree condition above.

Finally, an easy modification of the arguments at the end of the proof of the proposition shows that the conclusion still holds when H is a nef **R**-divisor.

PROOF OF THE PROPOSITION. Set $B = \{c_1, \ldots, c_b\}$. Let C' be the normalization of the image of f.

If C' is rational and f has degree $\geq b/2$ onto its image, just take $\Gamma = C'$. From now on, we will assume that if C' is rational, f has degree $< b/2$ onto its image.

By 2.9, the dimension of the space of morphisms from C to $f(C)$ that send B to $f(B)$ is at most $h^0(C, f^*T_{C'} \otimes I_B)$. When C' is irrational, $f^*T_{C'} \otimes I_B$ has negative degree, and, under our assumption, this remains true when C' is rational.

In both cases, the space is therefore 0-dimensional, hence any 1-dimensional subvariety of $\operatorname{Mor}(C, X; f|_B)$ through $[f]$ corresponds to morphisms with varying images.

Let $\overline{T}$ be a smooth compactification of the normalization of such a subvariety. Resolve the indeterminacies of the rational map $\operatorname{ev} : C \times \overline{T} \dashrightarrow X$ by blowing up points to get a morphism

$$e : S \xrightarrow{\varepsilon} C \times \overline{T} \xdashrightarrow{\operatorname{ev}} X$$

whose image is a *surface*.

For $i = 1, \ldots, b$, we denote by $E_{i,1}, \ldots, E_{i,n_i}$ the (effective) inverse images on S of the (-1)-exceptional curves that appear every time some point lying over $\{c_i\} \times \overline{T}$ is blown up. We have

$$E_{i,j} \cdot E_{i',j'} = -\delta_{i,i'}\delta_{j,j'}$$

The notation is summed up in the following picture.

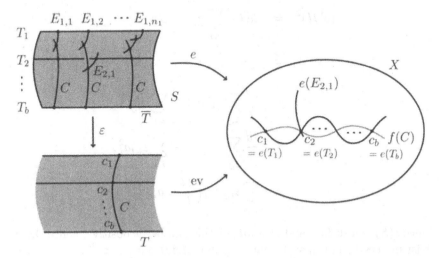

The curve $f(C)$ bends and breaks keeping $c_1, \ldots, c_b$ fixed.

Write the strict transform T_i of $\{c_i\} \times \overline{T}$ as

$$T_i \sim \varepsilon^*\overline{T} - \sum_{j=1}^{n_i} \varepsilon_{i,j} E_{i,j}$$

where $\varepsilon_{i,j} = T_i \cdot E_{i,j}$ is 1 if the blown up point is on the (smooth) strict transform of $\{c_i\} \times \overline{T}$, and 0 if it is not.

Write also

$$e^*H \sim a\varepsilon^*C + d\varepsilon^*\overline{T} - \sum_{i=1}^{b} \sum_{j=1}^{n_i} a_{i,j} E_{i,j} + G$$

where the $a_{i,j}$ are nonnegative and G is orthogonal to the **R**-vector subspace of $N^1(X)_{\mathbf{R}}$ generated by ε^*C, $\varepsilon^*\overline{T}$ and the $E_{i,j}$.

Note that e^*H is nef, hence

$$a = e^*H \cdot \varepsilon^*\overline{T} \geq 0$$

Since T_i is contracted by e to $f(c_i)$, we have for each i

$$0 = e^*H \cdot T_i = a - \sum_{j=1}^{n_i} \varepsilon_{i,j} a_{i,j}$$

Summing up over i, we get

$$ba = \sum_{i,j} \varepsilon_{i,j} a_{i,j} \tag{3.1}$$

Moreover, since $\varepsilon^* C \cdot G = 0 = (\varepsilon^* C)^2$ and $\varepsilon^* C$ is nonzero, the Hodge index theorem ([H1], V, rem. 1.9.1) implies $G^2 \leq 0$, hence (using (3.1))

$$
\begin{aligned}
(e^* H)^2 &= 2ad - \sum_{i,j} a_{i,j}^2 + G^2 \\
&\leq 2ad - \sum_{i,j} a_{i,j}^2 \\
&= \frac{2d}{b} \sum_{i,j} \varepsilon_{i,j} a_{i,j} - \sum_{i,j} a_{i,j}^2 \\
&\leq \frac{2d}{b} \sum_{i,j} \varepsilon_{i,j} a_{i,j} - \sum_{i,j} \varepsilon_{i,j} a_{i,j}^2 \\
&= \sum_{i,j} \varepsilon_{i,j} a_{i,j} \left(\frac{2d}{b} - a_{i,j} \right)
\end{aligned}
$$

Since $e(S)$ is a surface and H is ample, this number is positive, hence there exist indices i_0 and j_0 such that $\varepsilon_{i_0,j_0} > 0$ and $0 < a_{i_0,j_0} < \frac{2d}{b}$.

But $d = e^* H \cdot \varepsilon^* C = H \cdot C$, and $e^* H \cdot E_{i_0,j_0} = a_{i_0,j_0}$ is the H-degree of the rational 1-cycle $e_* E_{i_0,j_0}$. The latter is nonzero since $a_{i_0,j_0} > 0$, and it passes through $f(c_{i_0})$, since E_{i_0,j_0} meets T_{i_0} (their intersection number is $\varepsilon_{i_0,j_0} = 1$) and the latter is contracted by e to $f(c_{i_0})$. This proves the proposition: take for Γ a component of $e_* E_{i_0,j_0}$ which passes through $f(c_{i_0})$ but is not contracted by e. □

3.4   Rational Curves on Varieties Whose Canonical Divisor Is Not Nef

We proved in Section 3.2 that when X is a smooth projective variety such that $-K_X$ is ample (i.e., X is a Fano variety), there is a rational curve through any point of X. The following result considerably weakens the hypothesis: assuming only that K_X has negative degree on *one* curve C, we still prove that there is a rational curve through any point of C.

Note that the proof of Theorem 3.4 goes through in positive characteristic under this weaker hypothesis and does prove the existence of a rational curve through any point of C. However, to pass to the characteristic 0 case, one needs to bound the degree of this rational curve by some "universal" constant so that we deal only with a quasi-projective part of a morphism space. Apart from that, the ideas are essentially the same as in Theorem 3.4.

The following theorem first appeared in [MM] where it was presented as a criterion for uniruledness (see Definition 4.1 for the meaning of this term). We will present an application to varieties with nef anticanonical divisor (Theorem 3.10). It will also be used to prove the cone theorem in the smooth case (see Chapter 6) and Kawamata's theorem on lengths of rational curves (Theorem 7.46).

Theorem 3.6 *Let X be a projective variety, let H be an ample divisor on X, and let $f : C \to X$ be a smooth curve such that X is smooth along $f(C)$ and $K_X \cdot C < 0$. Given any point x on $f(C)$, there exists a rational curve Γ on X through x with*

$$H \cdot \Gamma \leq 2 \dim(X) \, \frac{H \cdot C}{-K_X \cdot C}$$

When X is smooth, the rational curve can be broken up, using Proposition 3.2 and (2.2), into several pieces (of lower H-degree) keeping any two points fixed (one of which being on $f(C)$), until one gets a rational curve Γ which satisfies $-K_X \cdot \Gamma \leq \dim(X) + 1$, in addition to the bound on its H-degree given above.

It is nevertheless useful to have a more general statement allowing X to be singular. It shows that a normal projective variety X with ample (**Q**-Cartier) anticanonical divisor is covered by rational curves of $(-K_X)$-degree at most $2\dim(X)$, and will be used in this form in Section 7.11.

Again, the theorem still holds when H is a nef **R**-divisor, but this is not as easy to prove as in Proposition 3.5 (see [K1], th. II.5.8).

PROOF OF THE THEOREM. The idea is to take b as big as possible in Proposition 3.5, in order to get the lowest possible degree for the rational curve. As in the proof of Theorem 3.4, we first assume that the characteristic p of the ground field **k** is positive, and use the Frobenius morphism to construct sufficiently many morphisms from C to X.

Assume then $p > 0$. We compose f with m Frobenius morphisms to get $f_m : C_m \to X$ of degree $p^m \deg(f)$ onto its image. We have by 2.11

$$\dim_{[f_m]} \mathrm{Mor}(C_m, X; f_m|_{B_m}) \geq -p^m K_X \cdot C + (1 - g(C) - b_m) \dim(X)$$

which is positive if we take

$$b_m = \left[\frac{-p^m K_X \cdot C}{\dim(X)} - g(C) \right]$$

which is a positive number for m sufficiently large. This is what we need to apply Proposition 3.5. It follows that there exists a rational curve Γ_m through some point of $f(B_m)$, such that

$$H \cdot \Gamma_m \leq \frac{2H \cdot C_m}{b_m} = \frac{2p^m}{b_m}(H \cdot C)$$

As m goes to infinity, p^m/b_m goes to $\dim(X)/(-K_X \cdot C)$. Since the left-hand side is an integer, we get

$$H \cdot \Gamma_m \leq \frac{2 \dim(X)}{-K_X \cdot C}(H \cdot C)$$

for m sufficiently large. By the lemma below, the set of points of $f(C)$ through which passes a rational curve of degree at most $2 \dim(X)\frac{H \cdot C}{-K_X \cdot C}$ is *closed* (it is the intersection of $f(C)$ and the image of the evaluation map). It cannot be finite since we could then take B_m such that $f_m(B_m)$ lies outside of that locus; hence it is equal to $f(C)$. This finishes the proof when the characteristic is positive.

As in the proof of Theorem 3.4, the characteristic 0 case is done by considering a finitely generated domain R over which X, C, f, H, and a point x of $f(C)$ are defined. The quasi-projective family of rational curves mapping 0 to x and of H-degree at most $2 \dim(X)\frac{H \cdot C}{-K_X \cdot C}$ is nonempty modulo any maximal ideal, hence is nonempty over the algebraic closure in **k** of the quotient field of R. $\square$

Lemma 3.7 *Let X be a projective variety and let d be a positive integer. Let M_d be the quasi-projective scheme that parametrizes rational curves on X of degree at most d. The image of the evaluation map*

$$\mathrm{ev}_d : \mathbf{P}^1 \times M_d \to X$$

is closed in X.

The image of ev_d is the set of points of X through which passes a rational curve of degree at most d.

PROOF OF THE LEMMA. The idea is that a rational curve can only degenerate into a union of rational curves of lower degrees.

Let x be a point in $\overline{\mathrm{ev}_d(M_d)} - \mathrm{ev}_d(M_d)$ and let T be the normalization of a 1-dimensional subvariety of $\mathbf{P}^1 \times M_d$ that dominates a curve in $\overline{\mathrm{ev}_d(M_d)}$ that passes through x and meets $\mathrm{ev}_d(M_d)$. Note that since x is not in the image of ev_d, the variety T is not contracted by the projection $\mathbf{P}^1 \times M_d \to M_d$. Let $\overline{T}$ be a smooth compactification of T.

The image of the rational map

$$\mathrm{ev} : \mathbf{P}^1 \times \overline{T} \dashrightarrow X$$

coming from the nonconstant morphism $T \to M_d$ is a surface and its indeterminacies can be resolved by blowing up a finite number of points to get a morphism

$$e : S \xrightarrow{\epsilon} \mathbf{P}^1 \times \overline{T} \xrightarrow{\mathrm{ev}} X$$

The surface $e(S)$ contains x. It is covered by the images of the fibers of the projection $S \to \overline{T}$, which are unions of rational curves of degree at most d. This proves the lemma. $\square$

Our next result generalizes Theorem 3.4 and shows that varieties with nef, but not numerically trivial, anticanonical divisor, are also covered by rational curves. One should be aware that this class of varieties is much larger than the class of Fano varieties.

We actually prove a result that applies to a slightly more general situation. Following [P], we will say that a Cartier divisor D on a projective variety X of dimension n is *generically nef* if $D \cdot H_1 \cdots H_{n-1} \geq 0$ for all ample divisors $H_1, \ldots, H_{n-1}$ on X. A nef divisor is generically nef, and so is an effective divisor. Through any point of X, there exists a curve intersection of $n - 1$ general hyperplane sections through x, and a generically nef divisor has nonnegative degree on that curve, hence the terminology.

Although we will not use this terminology very often, we will say that the Cartier divisor D is *generically ample* if $D \cdot H_1 \cdots H_{n-1} > 0$ for all ample divisors $H_1, \ldots, H_{n-1}$ on X.

3.8. Assume $D \cdot H^{n-1} = 0$ for all ample divisors H on X. For all t rational small enough, the Cartier divisor $H + tD$ is ample, hence $D \cdot (H + tD)^{n-1} = 0$, from which follows $D^2 \cdot H^{n-2} = 0$. *The Hodge index theorem implies that D is numerically trivial.* We prove it by induction on the dimension of X. The case $n = 2$ follows from the Hodge index theorem. Assume $n > 2$ and let C be an irreducible curve on X. Since H is ample, it follows from Bertini's theorem that for m sufficiently large, the zero set Y of a general element of $H^0(X, \mathscr{O}_X(mH))$ that vanishes on C is irreducible (use the same kind of argument as in footnote 12, p. 32, working on the blow-up of X along C). Since $H|_Y$ is ample and

$$D|_Y^i \cdot H|_Y^{n-1-i} = D^i \cdot H^{n-1-i} \cdot (mH) = 0$$

for $i = 1$ or 2, it follows by induction that $D|_Y$ is numerically trivial, hence $D \cdot C = D|_Y \cdot C = 0$.

There is a significant difference between nef and generically nef divisors, as shown by the following example.

Example 3.9 Let X be a $\mathbf{P}^1$-bundle over a smooth projective curve of genus g. The vector space $N^1(X)_{\mathbf{R}}$ is generated by the class of a fiber F and the class of a certain section C_0 ([H1], V, prop. 2.3). Assume $e = -C_0^2 \geq 0$.

A divisor class $a[C_0] + b[F]$ is nef (resp. ample) if and only if $a \geq 0$ and $b \geq ae$ (resp. $a > 0$ and $b > ae$) ([H1], V, prop. 2.20). Therefore, it is generically nef (resp. generically ample) if and only if $a \geq 0$ and $b \geq 0$ (resp. $a \geq 0$, $b \geq 0$ and $(a, b) \neq (0, 0)$).

Since $-K_X \sim 2C_0 + (2 - 2g + e)F$ ([H1], V, cor. 2.11), $-K_X$ is nef if and only if $g = 0$ and $e \leq 2$, or $g = 1$ and $e = 0$, whereas $-K_X$ is generically nef (or generically ample) if and only if $e \geq 2g - 2$ (any nonnegative value of e occurs for any given g by [H1], V, Example 2.11.3).

Theorem 3.10 *If X is a normal projective variety with $-K_X$ generically nef,*

- *either K_X is numerically trivial,[6]*

- *or there is a rational curve through any point of X.[7]*

Note that a general complete intersection curve on X does not meet the singular locus of X; hence it makes sense to say that $-K_X$ is generically nef.

PROOF OF THE THEOREM. Let $n = \dim(X)$. If $K_X \cdot H^{n-1} = 0$ for *all* ample divisors H, we saw in 3.8 that K_X is numerically trivial. Assume $K_X \cdot H^{n-1}$ is negative for *some* very ample divisor H. Let x be a smooth point of X and let C be the normalization of the intersection of $n - 1$ general hyperplane sections through x. As seen in footnote 12, p. 32, C is an irreducible curve and $K_X \cdot C = K_X \cdot H^{n-1} < 0$. By Theorem 3.6, there is a rational curve on X that passes through x (with the bound on its H-degree given in footnote 7, p. 70). One concludes with Lemma 3.7. $\square$

3.5   A Relative Bend-and-Break Lemma

We prove the first of two relative versions of the bend-and-break lemma: assume we have a morphism $\pi : X \to Y$ and a curve C on X that deforms nontrivially while fixing a point. Two things can happen:

- either the images in Y of the deformations of C are constant;

- or these images vary.

We deal here with the first case (the second case will be dealt with in Proposition 3.20) and prove that the rational curve produced by the standard bend-and-break Lemma 3.1 is contracted by π.

Proposition 3.11 *Let X and Y be projective varieties, let $\pi : X \to Y$ be a morphism, let $f : C \to X$ be an smooth irrational curve, and let c be a point on C. If $\dim_{[f]} \mathrm{Mor}_Y(C, X; f|_{\{c\}}) \geq 1$, there exist a curve $f' : C \to X$*

---

[6]If X is moreover smooth and the characteristic is zero, it is shown in [Ka1] that there exists a positive integer m such that $\mathscr{O}_X(mK_X) \simeq \mathscr{O}_X$. This also follows from the main result of [Si], which says that for each i, the locus $\{L \in \mathrm{Pic}^0(X) \mid H^i(X, L) \neq 0\}$ is a finite union of abelian subvarieties of $\mathrm{Pic}^0(X)$ translated *by torsion points* (apply with $i = \dim(X)$). In particular, there is a finite étale cover of X which has trivial canonical divisor. Beauville shows in [B1] that any projective complex variety with trivial canonical divisor has a finite étale cover which is a product of varieties of one of the three following types: abelian varieties, Calabi–Yau manifolds, and symplectic varieties (higher-dimensional analogs of K3 surfaces).

[7]Given an ample divisor H on X, one can also bound the H-degree by $2n \dfrac{H^n}{-K_X \cdot H^{n-1}}$, where $n = \dim(X)$. So X is *uniruled* in the terminology of Chapter 4.

and a connected nonzero effective rational 1-cycle Z on X passing through $f(c)$ such that

$$f_*C \sim f'_*C + Z , \quad \pi(Z) = \pi(f(c)) , \quad \pi \circ f' = \pi \circ f$$

This generalizes Proposition 3.1 (take for Y a point). According to 2.12 and the estimate (2.4), the hypotheses are fulfilled when X, Y, and C are smooth, π is surjective, $f(C)$ meets the locus where π is smooth, and

$$-K_{X/Y} \cdot f_*C - g(C)(\dim(X) - \dim(Y)) > 0$$

PROOF OF THE PROPOSITION. Let $\overline{T}$ be a smooth compactification of the normalization T of a 1-dimensional subvariety of $\mathrm{Mor}_Y(C, X; f|_{\{c\}})$ passing through $[f]$. The proof of Proposition 3.1 yields, with the same notation, a blow-up $\varepsilon : S \to C \times \overline{T}$ with an exceptional divisor E over some point (c, t_0) and a morphism $e : S \to X$. The morphism f' is the restriction of e to the strict transform of $C \times \{t_0\}$, the effective rational 1-cycle Z is e_*E, and all we have to do is to prove $\pi(e(E)) = \pi(f(c))$.

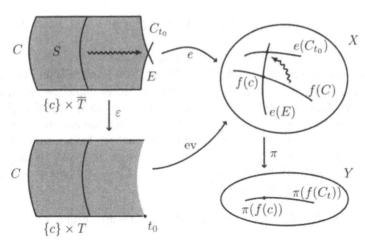

The curve $f(C)$ bends and breaks:
a vertical rational curve $e(E)$ appears.

Since $\pi \circ f = \pi \circ f_t$ for all t in T, we have for all c' in C

$$\pi \circ \mathrm{ev}(c', t) = \pi \circ f_t(c') = \pi \circ f(c') = \pi \circ f \circ p(c', t)$$

where p is the projection $C \times \overline{T} \to C$. It follows that $\pi \circ e$ and $\pi \circ f \circ p \circ \varepsilon$ coincide on $\varepsilon^{-1}(C \times T)$, hence on S. Since E is contracted by ε, the cycle $e(E)$ is contracted by π. $\qquad\square$

Using again reduction to positive characteristic, we deduce from the last result that given a morphism $\pi : X \to Y$ between smooth varieties, the relative anticanonical divisor $-K_{X/Y}$ is not ample, where

$$K_{X/Y} = K_X - \pi^* K_Y$$

Theorem 3.12 *Let X and Y be smooth projective varieties and let $\pi : X \to Y$ be a nonconstant generically smooth surjective morphism. Let H be an ample divisor on Y. For any positive ε, the divisor $-K_{X/Y} - \varepsilon\pi^* H$ is not nef. In particular, $-K_{X/Y}$ is not ample.*

By "generically smooth," it is meant that there exists a dense open subset Y_0 of Y such that the morphism $\pi^{-1}(Y_0) \to Y_0$ induced by π is smooth. This condition is automatic in characteristic zero.

PROOF OF THE THEOREM.[8] Assume that there exists $\varepsilon > 0$ such that $-K_{X/Y} - \varepsilon\pi^* H$ is nef. Let $f : C \to X$ be a smooth curve meeting $\pi^{-1}(Y_0)$ and not contracted by π, and let c be a point on C. Pick an ample divisor H_X on X and a positive rational α such that

$$(\varepsilon\pi^* H - \alpha H_X) \cdot f_* C > 0 \qquad (3.2)$$

Note that $-K_{X/Y} - \varepsilon\pi^* H + \alpha H_X$ is ample (1.22).

If the characteristic is 0, let R be a finitely generated domain over which X, H_X, Y, H, C, c, π and f are defined. Since ampleness is an open property ([Gr4], cor. 9.6.4) there exists a maximal ideal $\mathfrak{m}$ of R such that $-K_{X/Y} - \varepsilon\pi^* H + \alpha H_X$ is ample on the reduction modulo $\mathfrak{m}$. So we may assume that the characteristic p is positive, that we have a smooth curve $f : C \to X$ satisfying (3.2), and that $-K_{X/Y} - \varepsilon\pi^* H + \alpha H_X$ is ample.

By equation (2.4), for *any* smooth curve $f' : C' \to X$ whose image meets $\pi^{-1}(Y_0)$, and any point c' on C', we have

$$\dim_{[f']} \mathrm{Mor}_Y(C', X; f'|_{\{c'\}}) \geq -\deg(f'^* K_{X/Y}) - g(C')(\dim(X) - \dim(Y))$$

Since $-K_{X/Y} - \varepsilon\pi^* H + \alpha H_X$ is ample, this implies

$$\dim_{[f']} \mathrm{Mor}_Y(C', X; f'|_{\{c'\}}) > \qquad (3.3)$$
$$\deg(f'^*(\varepsilon\pi^* H - \alpha H_X)) - g(C')(\dim(X) - \dim(Y))$$

Let us go back to our curve $f : C \to X$. Because of (3.2), there exists a positive integer m such that

$$p^m \deg(f^*(\varepsilon\pi^* H - \alpha H_X)) - g(C)(\dim(X) - \dim(Y)) \geq 0$$

---

[8]The proof of this result in [Z] is slightly incorrect. The following argument was kindly provided by Q. Zhang.

Let $f_m : C_m \to X$ be the composition of f with m Frobenius morphisms. For any point c_m on C_m, we get by (3.3)

$$\dim_{[f_m]} \text{Mor}_Y(C_m, X; (f_m)|_{\{c_m\}}) \geq 1 \tag{3.4}$$

and by Proposition 3.11, there exists another morphism $f'_m : C_m \to X$ such that

$$\deg(f'^*_m H_X) < \deg(f^*_m H_X) , \qquad \pi \circ f'_m = \pi \circ f_m$$

We have then

$$
\begin{aligned}
\deg(f'^*_m(\varepsilon\pi^* H - \alpha H_X)) &= \varepsilon \deg((\pi \circ f'_m)^* H) - \alpha \deg(f'^*_m H_X) \\
&> \varepsilon \deg((\pi \circ f_m)^* H) - \alpha \deg(f^*_m H_X) \\
&\geq g(C)(\dim(X) - \dim(Y))
\end{aligned}
$$

Since $\pi \circ f'_m = \pi \circ f_m$, the image of f'_m still meets $\pi^{-1}(Y_0)$, so we still have (3.3) for f'_m, hence (3.4) is still valid for f'_m. This construction yields an infinite sequence of morphisms $C_m \to X$ with decreasing H_X-degrees. This is absurd and the theorem is proved. $\qquad\square$

3.6   Images of Varieties with Nef or Ample Anticanonical Divisor

We now use the results of the last section to examine to what extent positivity properties of the anticanonical divisor are preserved under a surjective morphism $\pi : X \to Y$ of smooth projective varieties. Things are not as nice as one could have hoped: even if X is a Fano variety, $-K_Y$ will not in general be nef (see Example 3.16(2)), but only generically ample (Corollary 3.14). However, if π is smooth, or if Y has Picard number 1, then Y is also a Fano variety (Corollary 3.15).

These results are based on the following proposition.

Proposition 3.13 *Let X and Y be projective varieties. Let $\pi : X \to Y$ be a surjective morphism and let $f : C \to Y$ be a smooth curve. Assume that X is smooth, that Y is smooth along $f(C)$, and that $X \times_Y C$ is smooth and generically smooth over C.[9] If D is a Cartier divisor on Y such that $-K_X + \pi^* D$ has nonnegative degree over any curve on X that dominates $f(C)$,*

$$(-K_Y + D) \cdot C \geq 0$$

---

[9]This last condition is automatic in characteristic zero.

PROOF. Set $X_C = X \times_Y C$. The second projection $\pi' : X_C \to C$ is a generically smooth surjective morphism between smooth varieties and

$$K_{X_C/C} = g^* K_{X/Y}$$

where $g : X_C \to X$ is the first projection.[10]

Let H be an ample divisor on Y and let ε be a positive number. By Theorem 3.12, $-K_{X_C/C} - \varepsilon \pi'^* f^* H$ is not nef hence there exists a curve $f' : C' \to X_C$ such that

$$
\begin{aligned}
0 \; > \; & (-K_{X_C/C} - \varepsilon \pi'^* f^* H) \cdot f'_* C' \\
= \; & (-g^* K_{X/Y} - \varepsilon g^* \pi^* H) \cdot f'_* C' \\
= \; & (-K_X + \pi^* K_Y - \varepsilon \pi^* H) \cdot (g \circ f')_* C'
\end{aligned}
$$

In particular, C' is not contracted by $g \circ f'$, hence maps onto a curve in X which dominates $f(C)$. By assumption, $(-K_X + \pi^* D) \cdot g(f'(C'))$ is nonnegative, hence

$$
\begin{aligned}
0 \; > \; & (-\pi^* D + \pi^* K_Y - \varepsilon \pi^* H) \cdot g(f'(C')) \\
= \; & (-D + K_Y - \varepsilon H) \cdot \pi_*(g(f'(C')))
\end{aligned}
$$

Since $\pi_*(g(f'(C')))$ is a positive multiple of C, we get

$$(-D + K_Y - \varepsilon H) \cdot C < 0$$

for all $\varepsilon > 0$, hence $(-D + K_Y) \cdot C \le 0$. $\qquad\square$

Corollary 3.14 *Assume the characteristic is zero. Let X and Y be projective varieties, with X smooth and Y nonsingular in codimension 1, let $\pi : X \to Y$ be a surjective morphism, and let D be a Cartier divisor on Y.*

(a) *If $-K_X + \pi^* D$ is nef, $-K_Y + D$ is generically nef.*

(b) *If $-K_X + \pi^* D$ is ample, $-K_Y + D$ is generically ample.*

PROOF. Let n be the dimension of Y and let $H_1, \ldots, H_{n-1}$ be ample divisors on Y. A general complete intersection curve C in Y of suitable multiples of $H_1, \ldots, H_{n-1}$ is smooth and connected by Bertini's theorem and so is $\pi^{-1}(C)$ because the characteristic is zero ([H1], III, Rem. 10.9.1).

---

[10]By [H3], rem. 1, p. 141, this base change formula holds for smooth or local intersection morphisms, but also, by the remark on pp. 143–144, for any composition of those. Note then that π is the composition of

$$\iota = (\mathrm{Id}_X, \pi) : X \hookrightarrow X \times Y$$

and of the projection $q : X \times Y \to Y$. If Y is smooth (which we can always assume upon replacing Y with its smooth locus), ι is a local complete intersection ([H3], Lemma 1.4, p. 141) and q is smooth since X is.

If $-K_X + \pi^*D$ is nef, the proposition yields

$$0 \le (-K_Y + D) \cdot C = (-K_Y + D) \cdot H_1 \cdots H_{n-1}$$

hence $-K_Y + D$ is generically nef. This proves (a).

If $-K_X + \pi^*D$ is ample, let ε be a positive rational number such that $-K_X + \pi^*D - \varepsilon\pi^*H_1$ is ample (1.20). We just proved

$$(-K_Y + D - \varepsilon H_1) \cdot H_1 \cdots H_{n-1} \ge 0$$

hence $-K_Y + D$ is generically ample. This proves (b).   □

This corollary fails in positive characteristic: Ekedahl constructs in [Ek] a smooth K3 surface over a field of characteristic 2 which is an (inseparable) double cover of a surface of general type.[11]

Corollary 3.15 *Let X and Y be smooth projective varieties, let $\pi : X \to Y$ be a morphism, and let D be a divisor on Y. Assume that either π is smooth, or $N^1(Y)_{\mathbf{R}}$ has dimension 1 and the characteristic is zero.*

(a) *If $-K_X + \pi^*D$ is nef, so is $-K_Y + D$.*

(b) *If $-K_X + \pi^*D$ is ample, so is $-K_Y + D$.*

PROOF. When $N^1(Y)_{\mathbf{R}}$ has dimension 1, a generically nef divisor is nef, and a generically ample divisor is ample. The corollary follows from Corollary 3.14 in this case.

When π is smooth, the hypotheses of Proposition 3.13 hold for any smooth curve on Y, and the same proof as in Corollary 3.14 applies.   □

Examples 3.16 (1) It follows from Example 3.9 that the assumption in Corollary 3.14(a) cannot be weakened to "$-K_X + \pi^*D$ generically nef" (and similarly for (b)): for any curve C, there is a ruled surface X over C with $-K_X$ generically ample, although $-K_C$ is in general very far from being ample.

(2) We can do even better. Consider the smooth projective variety $Y_{r,s}$ constructed in 1.36 and its blow-up $X_{r,s} \to Y_{r,s}$. Keeping the same notation, we have

$$[K_{Y_{r,s}}] = -(r+2)\xi + (r-s)h$$

The intersection with ℓ is $r - s$, hence $-K_{Y_{r,s}}$ is not nef when $r > s > 0$. On the other hand,

$$[K_{X_{r,s}}] = -(r+2)\xi + (r-s)h + r[E]$$

---

[11]Recall that a smooth projective variety X of dimension n is *of general type* if the canonical divisor K_X is *big* in the sense of 1.30, that is if

$$\liminf_{m \to +\infty} \frac{h^0(X, mK_X)}{m^n} > 0$$

hence

$$K_{X_{r\cdot s}} \cdot \ell = -s \quad , \quad K_{X_{r\cdot s}} \cdot \ell' = -2 \quad , \quad K_{X_{r\cdot s}} \cdot \ell'' = -r$$

so that, by (1.10) and Kleiman's criterion 1.27(a), $X_{r\cdot s}$ is a Fano variety.[12] *So we have an example of a Fano variety with a (smooth) image whose anticanonical divisor is not nef.* This shows that one cannot improve the conclusions of Corollary 3.14, and that the hypothesis "π smooth" cannot be dropped altogether in Corollary 3.15.

(3) Wiśniewski shows in [W] that Corollary 3.15(b) does not hold in general even if π is flat. His construction goes as follows. Let a and b be positive integers, set $r = 2a + 2b - 1$, and let $\mathscr{E}$ be the vector bundle on $\mathbf{P}^r$ associated with the locally free sheaf $\mathscr{O}_{\mathbf{P}^r} \oplus \mathscr{O}_{\mathbf{P}^r}(2a) \oplus \mathscr{O}_{\mathbf{P}^r}(2b)$. Let[13] $Y = \mathbf{P}(\mathscr{E})$ and let P be the image of the section of $Y \to \mathbf{P}^r$ corresponding to the trivial quotient of $\mathscr{E}$. The vector space $N^1(Y)_{\mathbf{R}}$ has dimension 2 and is generated by the class of a divisor D associated with the line bundle $\mathscr{O}_Y(1)$ and the class of the inverse image H of a hyperplane in $\mathbf{P}^r$. Moreover, $-K_Y \equiv 3D$ is not ample because it is trivial on P. It is however *nef*. Let

$$Z = \mathbf{P}(\mathscr{O}_Y(D + aH) \oplus \mathscr{O}_Y(D + bH) \oplus \mathscr{O}_Y(D + (a + b)H)) \xrightarrow{\ \pi\ } Y$$

and let G be a divisor associated with the *ample* line bundle $\mathscr{O}_Z(1)$. Wiśniewski checks that the linear system $|2G - 2(a + b)H|$ on Z contains a smooth element X (this is tricky, because it has a nonempty base-locus). Since $K_X \equiv K_Z + 2G - 2(a + b)H \equiv -G$, the variety X is a Fano variety. The morphism $X \to Y$ is flat (it is called a *conic bundle*), but Y is not a Fano variety.[14]

3.7 The Albanese Map of Varieties with Nef Anticanonical Divisor

We study here morphisms from a smooth projective variety X with nef anticanonical divisor to an abelian variety,[15] in characteristic zero. It is useful (but not essential) in this context to introduce the *Albanese variety* $\mathrm{Alb}(X)$ of X. It is an abelian variety with a morphism

$$\alpha_X : X \to \mathrm{Alb}(X)$$

---

[12]This can be seen more quickly through the description of $X_{r\cdot s}$ as a $\mathbf{P}^1$-bundle over $\mathbf{P}^r \times \mathbf{P}^s$ (see 1.36), using Example 3.3(2).

[13]As usual, we follow Grothendieck's notation: for a vector bundle $\mathscr{E}$, the projectivization $\mathbf{P}(\mathscr{E})$ is the space of *hyperplanes* in the fibers of $\mathscr{E}$.

[14]Note that the dimension $2a + 2b + 1$ of Y is at least 5. Wiśniewski's construction does not seem able to provide an example of a conic bundle $X \to Y$ with $-K_X$ ample but $-K_Y$ not nef.

[15]Recall that an abelian variety is a proper variety with a group law such that the multiplication and the inverse map are both morphisms; see [Mu1] for more details.

(the Albanese morphism) which is universal with respect to morphisms from X to an abelian variety: any such morphism factors through α_X. Over the complex numbers, the abelian variety $\mathrm{Alb}(X)$ can be constructed by transcendental techniques, using Hodge theory, as the cokernel of the group morphism

$$
\begin{aligned}
H_1(X, \mathbf{Z}) &\rightarrow H^0(X, \Omega^1_X)^* \\
\gamma &\mapsto \left(\omega \mapsto \int_\gamma \omega\right)
\end{aligned}
\tag{3.5}
$$

The Albanese morphism, which depends on the choice of a point x_0 in X, is given by

$$
\begin{aligned}
X &\rightarrow H^0(X, \Omega^1_X)^* / H_1(X, \mathbf{Z}) \\
x &\mapsto \left(\omega \mapsto \int_{x_0}^x \omega\right)
\end{aligned}
$$

The construction can be done in general in a purely algebraic setting ([Mu2], lecture 27). It follows from the universal property that the image of the Albanese morphism generates the group $\mathrm{Alb}(X)$.

Theorem 3.17 *Assume the characteristic is zero. The Albanese morphism of a smooth projective variety with nef anticanonical divisor is surjective and its fibers are connected.*

The conclusion can be stated without any reference to the Albanese variety (nor does the proof use the existence of the Albanese): in characteristic zero, the image of a morphism from a smooth projective variety with nef anticanonical divisor to an abelian variety is a (translated) abelian subvariety.

PROOF OF THE THEOREM. Let X be a smooth projective variety with nef anticanonical divisor. Consider the Stein factorization

$$
\alpha_X : X \xrightarrow{\tilde{\alpha}_X} Y \xrightarrow{\pi} \mathrm{Alb}(X)
$$

of the Albanese morphism, where $\tilde{\alpha}_X$ has connected fibers, Y is normal, and π is finite. We know from Corollary 3.14(a) that $-K_Y$ is generically nef, and the theorem follows from the next lemma. □

Lemma 3.18 *Assume the characteristic is zero. Let Y be a normal variety, let A be an abelian variety, and let $\pi : Y \rightarrow A$ be a finite map. If $-K_Y$ is generically nef, Y is an abelian variety.*

PROOF. Tangent spaces at various points of A can be identified by translation to the tangent space T at the origin. Let G the Grassmannian of n-dimensional vector subspaces of T, where n is the dimension of Y. The *Gauss map*

$$
\begin{aligned}
Y &\dashrightarrow G \\
y &\longmapsto T\pi(T_{Y,y})
\end{aligned}
$$

defines a *morphism* γ on some open subset Y_0 of the smooth locus of Y whose complement has codimension at least 2 (see 1.39). Consider the tautological rank n bundle S on G whose fiber at a point of G is the corresponding n-dimensional vector subspace of T. The tangent map

$$T\pi : T_{Y_0} \to \pi^* T_A \simeq \mathcal{O}_{Y_0} \otimes T$$

induces a generically bijective morphism $T_{Y_0} \to \gamma^* S$. The determinant $\wedge^n S$ being the dual of the ample line bundle $\mathcal{O}_G(1)$ coming from the Plücker embedding $G \subset \mathbf{P}(\wedge^n T)$, there is a generically bijective, hence injective morphism

$$\mathcal{O}_{Y_0}(-K_{Y_0}) \to \gamma^* \mathcal{O}_G(-1)$$

of line bundles on Y_0.

If $-K_Y$ is generically nef, so is $\gamma^* \mathcal{O}_G(-1)$. The dual of the latter being nef, they must be both numerically trivial by 3.8.

The first consequence is that γ is *constant*. This implies that the tangent map to the addition map $Y \times Y \to A$ has general rank n, hence, by generic smoothness, that $\pi(Y) + \pi(Y)$ has dimension n (see [H1], III, prop. 10.6): it must be a translate of $\pi(Y)$. A translate of $\pi(Y)$ is therefore a subgroup of A, hence an abelian subvariety. It is in particular smooth.

The second consequence is that by 1.41, the ramification divisor of the morphism $Y_0 \to \pi(Y)$ induced by π must be numerically trivial, hence must vanish. Since $Y - Y_0$ has codimension at least 2, π must be unramified by purity of the branch locus ([Gr6], Exp. X, th. 3.4(i)). A theorem of Serre and Lang ([Mu1], chap. IV, §18, p. 167) then implies that Y is an abelian variety. $\square$

Corollary 3.19 *With the hypotheses and notation of the theorem, we have*

$$h^1(X, \mathcal{O}_X) \le \dim(X)$$

Equality holds if and only if X is an abelian variety.

PROOF. The inequality comes from the fact that the dimension of $\mathrm{Alb}(X)$ is $h^0(X, \Omega_X^1) = h^1(X, \mathcal{O}_X)$ (see (3.5)). If equality holds, since $K_{\mathrm{Alb}(X)}$ is trivial, K_X is linearly equivalent to an effective divisor $\mathrm{Ram}(\alpha_X)$ whose support is the locus where α_X is ramified (1.41). But $-\mathrm{Ram}(\alpha_X)$ must then be nef, which can only happen if it is trivial. Hence α_X is unramified and X is an abelian variety. $\square$

Let again X be a smooth projective variety with nef anticanonical divisor. According to Theorem 3.10, either K_X is numerically trivial and we can apply footnote 6, p. 70, or X is covered by rational curves. Since any rational curve on X is contracted by α_X, we have

- either there exist a smooth projective variety Y with trivial canonical divisor and $H^1(Y, \mathscr{O}_Y) = 0$, an abelian variety A, and a Cartesian diagram

$$
\begin{array}{ccc}
A \times Y & \xrightarrow{\ pr_1\ } & A \\
\downarrow{\scriptstyle \pi} & & \downarrow{\scriptstyle \pi'} \\
X & \xrightarrow{\ \alpha_X\ } & \mathrm{Alb}(X)
\end{array}
$$

where π and π' are finite étale covers (in other words, α_X is a fiber bundle for the étale topology, with fiber Y),

- or any fiber of α_X is connected and covered by rational curves (it is *uniruled* in the terminology of Chapter 4).

The Albanese variety of a Fano variety is trivial: this follows from Theorem 3.17 and Corollary 3.14(b),[16] so Fano varieties obviously fall into the second category.

3.8   Another Relative Bend-and-Break Lemma

We now state and prove our second relative version of the bend-and-break lemma: the case where we are given a morphism $\pi : X \to Y$ and a curve C on X which deforms nontrivially while fixing a point, *with varying images on Y*. We prove that the rational curve produced by the standard bend-and-break lemma 3.1 is not contracted by π.

Proposition 3.20 *Let X and Y be projective varieties, let $\pi : X \to Y$ be a morphism, let $f : C \to X$ be a smooth curve, and let c be a point on C. If the components of $\mathrm{Mor}(C, X; f|_{\{c\}})$ containing $[f]$ are not all contracted by the map*

$$
\rho : \mathrm{Mor}(C, X; f|_{\{c\}}) \to \mathrm{Mor}(C, Y; (\pi \circ f)|_{\{c\}})
$$

given by composition by π, there exists a connected effective rational 1-cycle on X through $f(c)$ on which π is nonconstant.

This result also generalizes Proposition 3.1 (take $X = Y$). The proof actually shows that there exist a morphism $f' : C \to X$ through $f(c)$ and a connected effective rational 1-cycle Z on X passing through $f(c)$ and not contracted by π such that

$$
f_* C \sim f'_* C + Z
$$

In particular, one component of Z meets $\pi^{-1}(\pi(x))$ and is not contracted by π.

---

[16]Or from the Kodaira vanishing theorem, which asserts that since $-K_X$ is ample, $h^1(X, \mathscr{O}_X) = h^{\dim(X)-1}(X, \mathscr{O}_X(K_X)) = 0$.

PROOF OF THE PROPOSITION. We may assume that C is irrational. Let $\overline{T}$ be a smooth compactification of the normalization T of a 1-dimensional subvariety of $\mathrm{Mor}(C, X; f|_{\{c\}})$ passing through $[f]$ and not contracted by ρ. With the notation of the proof of Proposition 3.11, all we have to do is to prove that $e(E)$ is not contracted by π.

Assume instead that all exceptional divisors above $\{c\} \times \overline{T}$ are contracted by $\pi \circ e$. This means (in the notation of the proof of Proposition 3.1) that $\pi \circ ev$ is defined at every point of, hence in a neighborhood of, $\{c\} \times \overline{T}$. As in the proof of Proposition 3.1, this implies that the image of $\pi \circ ev$ is a curve, which must be $\pi(f(C))$.

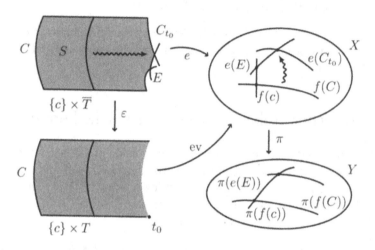

The curve $f(C)$ bends and breaks:
a nonvertical rational 1-cycle $e(E)$ appears.

Since C is irrational, this is possible only if the morphisms $\pi \circ f_t$ are all the same, and this contradicts the fact that $\rho(T)$ is a curve. $\qquad\square$

3.9 Transverse Rational Curves on Fano Varieties

We go back to a Fano variety X. At this point, we know that it is covered by rational curves. Our long-term aim is to show that any two points of X can be joined by a chain of rational curves (Proposition 5.16). This will require a lot of work, but the idea is very simple. We will first construct in general in Chapter 5 a kind of quotient $\rho : X \dashrightarrow Y$ for the equivalence relation "to be joined by a chain of rational curves," and then show that when X is a Fano variety, if Y is not a point, there must exist a rational curve on X not contracted by ρ, which is absurd. It is the existence of this "transverse" rational curve that we prove here, using the relative bend-and-break result proved in the last section, first assuming that ρ is a morphism, then in general.

Theorem 3.21 *Let X and Y be projective varieties and let $\pi : X \to Y$ be a nonconstant morphism. If X is a Fano variety, there exists, for any point y of $\pi(X)$, a rational curve on X of $(-K_X)$-degree at most $\dim(X) + 1$ which meets $\pi^{-1}(y)$ but is not contracted by π.*

Again, this result generalizes Theorem 3.4 (take $X = Y$).

PROOF OF THE THEOREM. We assume first that the characteristic of the ground field is positive and prove the existence of a transverse rational curve that meets $\pi^{-1}(y)$.

Since π is not constant, there exists a curve $f : C \to X$ not contracted by π that meets $\pi^{-1}(y)$. Let c be a point on C such that $\pi(f(c)) = y$. We may assume that C is irrational.

Fix an ample divisor H on Y. By 1.20, there exists a positive rational number α such that $-K_X - \alpha\pi^*H$ is ample on X. Since $\pi \circ f$ is not constant, we may assume, by composing f with sufficiently many Frobenius morphisms, that

$$\alpha H \cdot \pi_*(f_*C) \geq \dim(X)g(C) \tag{3.6}$$

It implies in particular

$$-K_X \cdot f_*C > \dim(X)g(C)$$

hence, by (2.2), that $\mathrm{Mor}(C, X; f|_{\{c\}})$ has positive dimension at $[f]$. If the conclusion of the theorem does not hold, Proposition 3.20 implies that each component of $\mathrm{Mor}(C, X; f|_{\{c\}})$ through $[f]$ is contracted by ρ. In that case, $\mathrm{Mor}_Y(C, X; f|_{\{c\}})$ has positive dimension at $[f]$ hence, by Proposition 3.11, there exist a curve $f' : C \to X$ and a connected nonzero effective rational 1-cycle Z on X passing through $f(c)$ such that

$$f_*C \sim f'_*C + Z , \qquad \pi \circ f' = \pi \circ f$$

from which it follows that the inequality (3.6) is still valid for f'. Also, since Z is nonzero,

$$0 < -K_X \cdot f'_*C < -K_X \cdot f_*C$$

We may repeat this construction infinitely many times, which is absurd since the $(-K_X)$-degrees of the curves would then form an infinite decreasing sequence of positive integers.

Once we know that there is a transverse rational curve meeting $\pi^{-1}(y)$, we can break it up fixing a point in $\pi^{-1}(y)$ and a point outside of it into rational curves of $(-K_X)$-degree at most $\dim(X)+1$ (Proposition 3.2), one of which must meet $\pi^{-1}(y)$ without being contained in it. This proves the theorem in positive characteristic.

The characteristic 0 case is treated by reduction to positive characteristic as in the proof of Theorem 3.4. All we have to do is to check that the space of maps whose existence we are trying to prove is quasi-projective. But, if $\rho : \mathrm{Mor}(\mathbf{P}^1, X) \to \mathrm{Mor}(\mathbf{P}^1, Y)$ is the map given by composition by π, this space can be defined, inside the quasi-projective variety $\mathrm{Mor}_{\leq \dim(X)+1}(\mathbf{P}^1, X)$, by the conditions $\rho(f)(0) = y$ (closed) and $\rho(f)(\infty) \neq y$ (open). It is therefore quasi-projective. $\square$

We now prove the following sharper form of the theorem, where π is only a rational map, which is, however, required to be proper on an open set.

Theorem 3.22 *Let X be a Fano variety and let Y_0 be a quasi-projective variety. Let X_0 be a dense open subset of X and let $\pi : X_0 \to Y_0$ be a proper nonconstant morphism. There exists, for any point y of $\pi(X_0)$, a rational curve on X that meets $\pi^{-1}(y)$ but is not contracted by π.*

PROOF. Let Y be a compactification of Y_0, let $\tilde{X}$ be the closure in $X \times Y$ of the graph of π, let $\varepsilon : \tilde{X} \to X$ be the birational morphism induced by the first projection, and let $\tilde{\pi} : \tilde{X} \to Y$ be the second projection.

Let H be an ample divisor on Y. Since X is smooth, there exists by 1.42 an effective $\mathbf{Q}$-divisor E with support contained in the exceptional locus of ε such that $\varepsilon^*(-K_X) - E$ is ample, and there exists a positive rational α such that $\varepsilon^*(-K_X) - E - \alpha\tilde{\pi}^*H$ is ample (1.20).

As above, we may assume that the ground field has positive characteristic, and start with a smooth curve $\tilde{f} : C \to \tilde{X}$ not contracted by $\tilde{\pi}$ and a point c such that $\tilde{f}(c) \in \varepsilon^{-1}(X_0)$ and $\tilde{\pi}(\tilde{f}(c)) = y$. Set $f = \varepsilon \circ \tilde{f}$.

Since $\tilde{\pi} \circ \tilde{f}$ is not constant, we may assume, by composing $\tilde{f}$ with sufficiently many Frobenius morphisms, that

$$\alpha H \cdot \tilde{\pi}_*(\tilde{f}_* C) \geq \dim(X) g(C) \tag{3.7}$$

It implies

$$
\begin{aligned}
-K_X \cdot f_* C &= -\varepsilon^* K_X \cdot \tilde{f}_* C \\
&> (E + \alpha\tilde{\pi}^* H) \cdot \tilde{f}_* C \\
&\geq \alpha\tilde{\pi}^* H \cdot \tilde{f}_* C \\
&\geq \dim(X) g(C)
\end{aligned}
$$

hence, by (2.2), there is a 1-dimensional subvariety of $\mathrm{Mor}(C, X; f|_{\{c\}})$ containing $[f]$. Let T be its normalization and let $\mathrm{ev} : C \times T \to X$ be the evaluation map. Its image meets X_0, so the composition $\varepsilon^{-1} \circ \mathrm{ev}$ defines a rational map $C \times T \dashrightarrow \tilde{X}$ whose indeterminacies can be resolved by blowing up finitely many points to get a morphism

$$e : S \xrightarrow{\varepsilon'} C \times T \xrightarrow{\mathrm{ev}} X \xrightarrow{\varepsilon^{-1}} \tilde{X}$$

such that $\varepsilon \circ e = \text{ev} \circ \varepsilon'$. Let T_0 be the complement in T of the projection of the (finite) set that ε' blows up. It might very well be that $[f]$ corresponds to a point of T that is not in T_0. The situation is summed up in, and I hope clarified by, the following commutative diagram

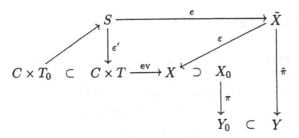

For any t in T, let C_t be the strict transform of $C \times \{t\}$ on S. We have $f_t = (\varepsilon \circ e)|_{C_t}$. Pick a point t_0 in T_0. The curve C_{t_0} is the fiber of t_0 for the projection $S \to T$ and $\tilde{f}_{t_0} = e|_{C_{t_0}}$. If t_1 is a point of T corresponding to $[f]$, we have $\tilde{f} = e|_{C_{t_1}}$ and there exists an effective rational 1-cycle Z on S such that $C_{t_0} \sim C_{t_1} + Z$.

There is a factorization

$$T_0 \to \mathrm{Mor}(C, \tilde{X}; \tilde{f}|_{\{c\}}) \to \mathrm{Mor}(C, X; f|_{\{c\}})$$

If the image of T_0 in $\mathrm{Mor}(C, \tilde{X}; \tilde{f}|_{\{c\}})$ is *not* contracted by the map

$$\rho : \mathrm{Mor}(C, \tilde{X}; \tilde{f}|_{\{c\}}) \to \mathrm{Mor}(C, Y; (\tilde{\pi} \circ \tilde{f})|_{\{c\}})$$

there exists by Proposition 3.20 a rational curve $\tilde{\Gamma}$ in $\tilde{X}$ that meets $\tilde{\pi}^{-1}(y)$ but is not contracted by $\tilde{\pi}$. Note that since π is proper, ε induces an isomorphism between $\tilde{\pi}^{-1}(y)$ and $\pi^{-1}(y)$. In particular, $\tilde{\Gamma}$ meets $\varepsilon^{-1}(X_0)$, hence $\varepsilon(\tilde{\Gamma})$ is not a point: this is the rational curve we are looking for.

Assume now that the image of T_0 is contracted by ρ. By Proposition 3.11, there exist a curve $\tilde{f}'_{t_0} : C \to \tilde{X}$ and a connected nonzero effective rational 1-cycle $\tilde{Z}$ on $\tilde{X}$ passing through $\tilde{f}_{t_0}(c) = \tilde{f}(c)$ such that

$$(\tilde{f}_{t_0})_* C \sim (\tilde{f}'_{t_0})_* C + \tilde{Z} \qquad \text{and} \qquad \tilde{\pi} \circ \tilde{f}'_{t_0} = \tilde{\pi} \circ \tilde{f}_{t_0}$$

The estimate (3.7) still holds for $\tilde{f}'_{t_0}$, because

$$
\begin{aligned}
H \cdot \tilde{\pi}_*((\tilde{f}'_{t_0})_* C) &= H \cdot (\tilde{\pi} \circ \tilde{f}'_{t_0})_* C = H \cdot (\tilde{\pi} \circ \tilde{f}_{t_0})_* C \\
&= H \cdot \tilde{\pi}_*(e_*(C_{t_0})) = H \cdot \tilde{\pi}_*(e_*(C_{t_1} + Z)) \\
&\geq H \cdot \tilde{\pi}_*(e_*(C_{t_1})) = H \cdot \tilde{\pi}_*(\tilde{f}_* C)
\end{aligned}
$$

On the other hand, since $\tilde{Z}$ meets $\varepsilon^{-1}(X_0)$ hence is not contracted by ε, we have

$$0 < -K_X \cdot (\varepsilon \circ \tilde{f}'_{t_0})_* C < -K_X \cdot (\varepsilon \circ \tilde{f}_{t_0})_* C = -K_X \cdot (f_{t_0})_* C = -K_X \cdot f_* C$$

We may repeat this construction infinitely many times, which is absurd since the $(-K_X)$-degrees of the curves would then form an infinite decreasing sequence of positive integers.   □

3.10   Exercises

1. **(M. Raynaud).** Let C be a smooth curve of positive genus defined over an algebraically closed field $\mathbf{k}$ of positive characteristic p and let $\mathscr{L}$ be an invertible sheaf of order p in $\mathrm{Pic}^0(C)$. To $\mathscr{L}$ is associated a purely inseparable cover $\pi : C' \to C$ by setting

$$C' = \mathbf{Spec}(\mathscr{O}_C \oplus \mathscr{L} \oplus \cdots \oplus \mathscr{L}^{p-1})$$

Alternatively, if $\mathscr{L}$ is given by transition functions g_{ij} with respect to an open affine cover (U_i) such that

$$g_{ij}^p = \frac{f_i}{f_j}, \qquad f_i \in \mathscr{O}(U_i)^*$$

the curve C' is given in $U_i \times \mathbf{k}$ by the equation $f_i = t^p$.

   (a) Show that the $\frac{df_i}{f_i}$ define a global regular nonzero 1-form ω on C.

   (b) Show that π is a homeomorphism and that C' is singular exactly over the points where ω vanishes, where it has a cusp.

   (c) Assume that C is embedded in an abelian variety A and that $\mathscr{L}$ is the restriction of an invertible sheaf on A, to which correspond an abelian variety A' with an inseparable isogeny $A' \to A$, and a commutative diagram

$$\begin{array}{ccc} C' & \hookrightarrow & A' \\ \downarrow{\scriptstyle\pi} & & \downarrow \\ C & \hookrightarrow & A \end{array}$$

   Show that the Gauss map of the subvariety C' of A' is constant, although C' is not an elliptic curve.

2. Let X be a smooth projective variety, let D be a hypersurface in X, and let C be a curve in X that does not meet the singular locus of D. If

$$K_X \cdot C = 0, \qquad D \cdot C < 0$$

prove that X contains a rational curve.[17]

---

[17]We will see in Exercise 7.13.8(c) that the conclusion holds even if C meets the singular locus of D.

4
Uniruled and Rationally Connected Varieties

As we saw in Section 3.2, there exists a rational curve through every point of a Fano variety. A variety X with this property is called *uniruled:* there exists on X a rational curve whose deformations cover a dense open subset of X. We show that most other versions of this definition that come to mind turn out to be equivalent, at least over an uncountable algebraically closed field: the thing that one wants to rule out is a variety not covered by rational curves say of fixed degree (with respect to some ample divisor), but which is still a (countable) union of rational curves (the degrees going to infinity).

We then restrict ourselves to *smooth* proper varieties. It turns out that *in characteristic zero* (this is false in positive characteristic), such a variety X is uniruled if and only if there exists on X a rational curve over which the tangent bundle T_X is generated by global sections (each rank 1 piece has nonnegative degree). Such a curve is called *free*. This surprising criterion is in fact very easy to prove: one needs to show that some evaluation map

$$\mathbf{P}^1 \times \mathrm{Mor}_d(\mathbf{P}^1, X) \to X$$

is dominant, and by generic smoothness, this is equivalent to its differential being surjective at some point of $\mathbf{P}^1 \times \mathrm{Mor}_d(\mathbf{P}^1, X)$. An explicit calculation shows that this is the case if and only if the corresponding rational curve $\mathbf{P}^1 \to X$ is free.

Free rational curves will also play an important role in Section 4.6, because chains of free rational curves are easy to smooth.

We then study an analogous (stronger) property with the same kind of geometric flavor. A proper variety on which a general pair of points can be connected by a rational curve is called *rationally connected*. We will prove in the next chapter that Fano varieties have this property. Here we just follow the same path as for uniruledness. The definition we just gave is actually valid only over an uncountable algebraically closed field, for the same reasons as above. Apart from that slight trap, everything works fine.

Again, in characteristic zero, a *smooth* proper variety X is rationally connected if and only if there exists on X a rational curve over which the tangent bundle T_X is ample (each rank 1 piece has positive degree). Such a curve is called *very free*. The proof is essentially the same as above.

These curves will also play an important role in Section 4.6, for similar reasons, the guiding idea being that a free rational curve moves freely, while a very free rational curve moves freely even with one point fixed.

The next property we study is also of a very geometric nature: a proper variety for which a pair of general points can be connected by a chain of rational curves is said to be *rationally chain-connected*. Our main result is that in *characteristic zero*, a *smooth* proper rationally chain-connected variety is rationally connected (this was originally proved in [KMM2]). Moreover, there is a very free rational curve passing through *any* given finite subset. Although the idea is simple (one needs to smooth a chain of rational curves), the actual proof is rather technical and involves difficult smoothing results for trees of rational curves proved in Section 4.6.

All varieties are defined over a field **k** which will be assumed to be algebraically closed as of Section 4.2.

4.1 Uniruled and Rationally Connected Varieties

As explained in the introduction, we explore various possible formalizations of the geometrically intuitive notions of uniruledness and rational connectedness and show that they are all equivalent over an uncountable algebraically closed field.

Definition 4.1 *A variety X of dimension n is called* uniruled *if there exist a variety Y of dimension $n-1$ and a dominant rational map $\mathbf{P}^1 \times Y \dashrightarrow X$.*

In the same way that a "unirational" variety is dominated by a rational variety, a "uniruled" variety is dominated by a ruled variety. Hence the terminology.

Here are various other characterizations and properties of uniruled varieties.

Remarks 4.2 (1) A point is not uniruled. Any variety birationally isomorphic to a uniruled variety is uniruled. Any unirational variety that is not a point is uniruled. The product of a uniruled variety with any variety is uniruled.

(2) A variety X of dimension n is uniruled if and only if there exist a variety Y, an open subset U of $\mathbf{P}^1 \times Y$, and a dominant map $e : U \to X$ such that for some y in Y, the set $U \cap (\mathbf{P}^1 \times \{y\})$ is nonempty and not contracted by e.

This can be seen as follows. We may assume that Y is affine of dimension $m \geq n$. The hypothesis says that some fiber of the morphism

$$\{(t, t', y) \in \mathbf{P}^1 \times \mathbf{P}^1 \times Y \mid (t, y), (t', y) \in U \text{ and } e(t, y) = e(t', y)\} \to \mathbf{P}^1 \times Y$$

defined by $F((t, y), (t', y)) = (t, y)$ is finite. By Chevalley's theorem ([H1], II, ex. 3.22), a general fiber of F is finite, which means that a general (t, y) in U is isolated in its fiber for the projection $q : U_{(t,y)} = e^{-1}(e(t, y)) \to Y$. By Chevalley's theorem again, $U_{(t,y)}$ has dimension $m + 1 - n$ at (t, y); hence so has $q(U_{(t,y)})$ at y. Since $m + 1 - n > 0$, a general hyperplane section H of Y passing through y is transverse to $q(U_{(t,y)})$ at y, hence the dimension of $q^{-1}(H)$ at (t, y) is $m - n$. But this is the dimension at (t, y) of the fiber of (t, y) for the restriction of e to $U \cap (\mathbf{P}^1 \times H)$. By Chevalley's theorem again, the image of this restriction has dimension at least $1 + m - 1 - (m - n) = n$, hence must be dense in X. We may continue the process as long as $m \geq n$, hence X is uniruled.

(3) Let X be a *proper* uniruled variety, with a dominant rational map $e : \mathbf{P}^1 \times Y \dashrightarrow X$ as in the definition. We may compactify Y then normalize it. The map e is then defined outside of a subvariety of $\mathbf{P}^1 \times Y$ of codimension at least 2 (1.39), which therefore projects onto a proper closed subset of Y. By shrinking Y, we may therefore assume that e is a *morphism*; if X is projective, it factors as

$$\mathbf{P}^1 \times Y \to \mathbf{P}^1 \times \mathrm{Mor}_d(\mathbf{P}^1, X) \xrightarrow{\mathrm{ev}_d} X$$

for some *positive* integer d, and ev_d is dominant.

(4) Assume $\mathbf{k}$ is algebraically closed. It follows from (3) that there is a rational curve through a general point of a proper uniruled variety (actually, by Lemma 3.7, there is even a rational curve through *every* point). The converse holds *if $\mathbf{k}$ is uncountable*.

Indeed, we may, after shrinking and compactifying X, assume that it is projective. There is still a rational curve through a general point, and this is exactly saying that the evaluation map $\mathrm{ev} : \mathbf{P}^1 \times \mathrm{Mor}_{>0}(\mathbf{P}^1, X) \to X$ is dominant. Since $\mathrm{Mor}_{>0}(\mathbf{P}^1, X)$ has at most countably many irreducible components and X is not the union of countably many proper subvarieties, the restriction of ev to at least one of these components must be surjective, hence X is uniruled by (2).

(5) If X is a variety defined over a field k and K is an extension of k, the variety X_k is uniruled if and only if X_K is uniruled.

Indeed, we may assume, after shrinking and compactifying X, that it is projective. If X_K is uniruled, there exists by (3) a positive integer d such that the evaluation map

$$\mathrm{ev}_d^K : \mathbf{P}_K^1 \times \mathrm{Mor}_d(\mathbf{P}_K^1, X_K) \to X_K$$

is dominant. Because of the universal property of the parameter space $\mathrm{Mor}_d(\mathbf{P}_k^1, X)$, the scheme $\mathrm{Mor}_d(\mathbf{P}_K^1, X_K)$ is its extension to K and ev_d^K the extension of ev_d^k. It follows that the map ev_d^K is also dominant.

Because of (4), it is often useful to replace the base field by an algebraically closed and uncountable extension.

(6) A connected finite étale cover of a proper uniruled variety is uniruled.

Indeed, let X be a proper uniruled variety and let $\pi : \tilde{X} \to X$ be a connected finite étale cover. We may assume by Chow's lemma (which says that there exist a projective variety X' and a surjective birational morphism $X' \to X$) that X is projective, in which case there exists by (3) a positive integer d such that ev_d is dominant. Since $\mathbf{P}^1$ is simply connected ([H1], IV, ex. 2.5.3), the morphism $\mathrm{Mor}_d(\mathbf{P}^1, \tilde{X}) \to \mathrm{Mor}_d(\mathbf{P}^1, X)$ given by composition by π is surjective. It follows that $\tilde{X}$ is uniruled.

Definition 4.3 *A variety X is called* rationally connected *if it is* proper[1] *and if there exist a variety M and a rational map $e : \mathbf{P}^1 \times M \dashrightarrow X$ such that the rational map*

$$
\begin{array}{ccc}
\mathbf{P}^1 \times \mathbf{P}^1 \times M & \dashrightarrow & X \times X \\
(t, t', z) & \longmapsto & (e(t, z), e(t', z))
\end{array}
$$

is dominant.

Remarks 4.4 (1) A point is rationally connected. Rational connectedness is a birational property (for proper varieties!). Better, if X is a rationally connected variety and $X \dashrightarrow Y$ a dominant rational map, Y is rationally connected. In particular, a proper unirational variety is rationally connected. A (finite) product of rationally connected varieties is rationally connected. A rationally connected variety is uniruled.

(2) In the definition, the rational map e may be assumed to be a morphism (proceed as in Remark 4.2(3)). If X is a projective rationally connected variety, this implies that the morphism

$$
\begin{array}{ccc}
\mathbf{P}^1 \times \mathbf{P}^1 \times \mathrm{Mor}_d(\mathbf{P}^1, X) & \to & X \times X \\
(t, t', f) & \mapsto & (f(t), f(t'))
\end{array}
$$

---

[1]One can make a similar definition for nonproper varieties, with a choice of requiring e to be either just a rational map as here, which makes rational connectedness a birational property, or a morphism, as in [K1], in which case a positive-dimensional affine variety is not rationally connected since it contains no rational curves. To avoid these subtleties, we will stick to the proper case.

is dominant for some d. If **k** is algebraically closed, the evaluation map

$$\mathbf{P}^1 \times \mathrm{Mor}_d(\mathbf{P}^1, X; 0 \mapsto x) \quad \rightarrow \quad X$$
$$(t, f) \qquad\qquad\qquad \mapsto \quad f(t)$$

is dominant for x general in X.

(3) Assume **k** is algebraically closed. On a rationally connected variety, a general pair of points can be joined by a rational curve.[2] Again, the converse holds *if* **k** *is uncountable* (with the same proof as in Remark 4.2(4)).

(4) If X is a variety defined over a field k and K is an extension of k, the variety X_k is rationally connected if and only if X_K is rationally connected (proceed as in Remark 4.2(5)).

(5) Any proper variety that is an étale cover of a rationally connected variety is rationally connected. The proof is analogous to the proof of Remark 4.2(6): we may assume that the rationally connected variety X is projective and that the evaluation map

$$\mathbf{P}^1 \times \mathbf{P}^1 \times \mathrm{Mor}_d(\mathbf{P}^1, X) \rightarrow X \times X$$

is dominant for some d. Again, if $\tilde{X}$ is a proper variety and $\pi : \tilde{X} \rightarrow X$ a connected étale cover, the morphism $\mathrm{Mor}_d(\mathbf{P}^1, \tilde{X}) \rightarrow \mathrm{Mor}_d(\mathbf{P}^1, X)$ given by composition by π is surjective. It follows that $\tilde{X}$ is rationally connected.

However, this is not very meaningful since we will see in Corollary 4.18 that any such cover of a smooth projective rationally connected variety is trivial, at least in characteristic zero.

As of now, we will assume that the ground field **k** is algebraically closed. When it is moreover uncountable, the definitions of uniruled and rationally connected varieties become very simple: a variety is uniruled if there is a rational curve through a general point. It is rationally connected if it is proper and there is a rational curve through a general pair of points.

4.2 Free Rational Curves and Uniruled Varieties

For a smooth projective variety X defined over an algebraically closed field of characteristic zero, we show in this section that uniruledness is equivalent to the existence of a single rational curve on which the tangent bundle T_X is generated by global sections. This will be proved in Corollary 4.11.

Let X be a smooth variety of dimension n and let $f : \mathbf{P}^1 \rightarrow X$ be a morphism. By Grothendieck's theorem ([H1], V, ex. 2.6), the vector bundle f^*T_X decomposes as a sum of line bundles

$$f^*T_X \simeq \mathscr{O}_{\mathbf{P}^1}(a_1) \oplus \cdots \oplus \mathscr{O}_{\mathbf{P}^1}(a_n) \tag{4.1}$$

---

[2]In characteristic zero, we will prove in Corollary 4.28 that *any* two points of a *smooth* projective rationally connected variety can be joined by a rational curve.

where we assume $a_1 \geq \cdots \geq a_n$. If f is nonconstant (i.e., when it is a rational curve), f^*T_X contains $T_{\mathbf{P}^1} \simeq \mathcal{O}_{\mathbf{P}^1}(2)$ and $a_1 \geq 2$.

If $H^1(\mathbf{P}^1, f^*T_X)$ vanishes, the space $\mathrm{Mor}(\mathbf{P}^1, X)$ is smooth at $[f]$ (Theorem 2.6). This happens exactly when $a_n \geq -1$.

We will see that the bigger a_n is, the more the curve can be moved. This will lead to a simple characterization of uniruled and rationally connected varieties in characteristic zero.

Note that f^*T_X is generated by its global sections if and only if $a_n \geq 0$.

Definition 4.5 *Let r be a nonnegative integer. A rational curve $f : \mathbf{P}^1 \to X$ on a smooth variety X is r-free if $f^*T_X \otimes \mathcal{O}_{\mathbf{P}^1}(-r)$ is generated by its global sections.*

With our notation, this means $a_n \geq r$. We will say "free" instead of "0-free," and "very free" instead of "1-free." For easier statements, we will also agree that a constant morphism $\mathbf{P}^1 \to X$ is never free, and is very free if and only if X is a point. Note that given a very free rational curve, its composition with a (ramified) finite map $\mathbf{P}^1 \to \mathbf{P}^1$ of degree r is r-free.

Remark 4.6 A rational curve $f : \mathbf{P}^1 \to X$ on a smooth variety X is r-free if and only if

$$H^1(\mathbf{P}^1, f^*T_X \otimes \mathcal{O}_{\mathbf{P}^1}(-r-1)) = 0$$

The semi-continuity theorem ([H1], III, th. 12.8) implies that the set of r-free rational curves on X is an open subset of $\mathrm{Mor}(\mathbf{P}^1, X)$, possibly empty, which is smooth by Theorem 2.6.

Examples 4.7 (1) If $f : \mathbf{P}^1 \to X$ is a free rational curve, $K_X \cdot f_*\mathbf{P}^1 = -\sum_{i=1}^n a_i \leq -2$. There are no free rational curves on a variety whose canonical divisor is nef.

(2) Any rational curve in $\mathbf{P}^n$ is very free. This is because $T_{\mathbf{P}^n}$ is ample (see [H2]), hence so is its pull-back by a finite morphism. Alternatively, $T_{\mathbf{P}^n}$ is a quotient of $\mathcal{O}_{\mathbf{P}^n}(1)^{n+1}$, hence its inverse image by f is a quotient of $\mathcal{O}_{\mathbf{P}^1}(d)^{n+1}$, where d is the degree of f. Each $\mathcal{O}_{\mathbf{P}^1}(a_i)$ is a quotient of $\mathcal{O}_{\mathbf{P}^1}(d)^{n+1}$, hence $a_i \geq d$.

(3) A *smooth* rational curve C on a smooth surface X is free (resp. very free) if and only if $C^2 \geq 0$ (resp. $C^2 > 0$). Indeed, there is an exact sequence

$$0 \to \mathcal{O}_C(2) \to T_X|_C \to \mathcal{O}_C(C^2) \to 0 \tag{4.2}$$

Write $T_X|_C \simeq \mathcal{O}_{\mathbf{P}^1}(a_1) \oplus \mathcal{O}_{\mathbf{P}^1}(a_2)$ with $a_1 \geq a_2$. Either

$$a_2 \geq 2 \quad \text{and} \quad C^2 = a_1 + a_2 - 2 > 0$$

or the sequence (4.2) is split and

$$\{a_1, a_2\} = \{2, C^2\}$$

In particular, if a_2 or C^2 are nonpositive, they are equal, hence $C^2 \geq 0$ (resp. $C^2 > 0$) if and only if $a_2 \geq 0$ (resp. $a_2 > 0$).

Informally speaking, the freer a rational curve is, the more it can move while keeping points fixed. The precise result is the following.

Proposition 4.8 *Let X be a smooth quasi-projective variety, let r be a nonnegative integer, let $f : \mathbf{P}^1 \to X$ be an r-free rational curve and let B be a finite subscheme of $\mathbf{P}^1$ of length b. Let s be a positive integer such that $s + b \leq r + 1$. The evaluation map*

$$\begin{array}{cccc} \mathrm{ev}: & (\mathbf{P}^1)^s \times \mathrm{Mor}(\mathbf{P}^1, X; f|_B) & \to & X^s \\ & (t_1, \ldots, t_s, g) & \mapsto & (g(t_1), \ldots, g(t_s)) \end{array}$$

is smooth at $(t_1, \ldots, t_s, f)$ when $\{t_1, \ldots, t_s\} \cap B = \varnothing$.

Geometrically speaking, the proposition means that the deformations of an r-free rational curve keeping b points fixed ($b \leq r$) pass through $r + 1 - b$ general points of X. In particular, deformations of a free rational curve cover a dense subset of X and so do deformations of a very free rational curve keeping one point fixed.

PROOF OF THE PROPOSITION. The tangent map to the morphism ev at the point $(t_1, \ldots, t_s, f)$ is the map

$$\bigoplus_{i=1}^{s} T_{\mathbf{P}^1, t_i} \oplus H^0(\mathbf{P}^1, f^*T_X(-B)) \to \bigoplus_{i=1}^{s} T_{X, f(t_i)} \simeq \bigoplus_{i=1}^{s} (f^*T_X)_{t_i}$$

given by

$$(u_1, \ldots, u_s, \sigma) \mapsto (T_{t_1} f(u_1) + \sigma(t_1), \ldots, T_{t_s} f(u_s) + \sigma(t_s))$$

It is obviously surjective if the evaluation map

$$H^0(\mathbf{P}^1, f^*T_X(-B)) \longrightarrow H^0(\mathbf{P}^1, f^*T_X) \longrightarrow \bigoplus_{i=1}^{s}(f^*T_X)_{t_i}$$

$$\sigma \longmapsto (\sigma(t_1), \ldots, \sigma(t_s))$$

is. With the notation (4.1), this is in turn the case if each map

$$H^0(\mathbf{P}^1, \mathscr{O}_{\mathbf{P}^1}(a_j - b)) \to H^0(\mathbf{P}^1, \mathscr{O}_{\mathbf{P}^1}(a_j)) \to \bigoplus_{i=1}^{s} \mathbf{k}_{t_i}$$

for $j = 1, \ldots, n$, is surjective, and this holds since $\{t_1, \ldots, t_s\} \cap B = \varnothing$ and $a_j - b \geq s - 1$.

On the other hand, since $a_j - b \geq -1$, the group $H^1(\mathbf{P}^1, f^*T_X(-B))$ vanishes, hence $\mathrm{Mor}(\mathbf{P}^1, X; f|_B)$ is smooth at $[f]$ (2.9). This implies that ev is smooth at $(t_1, \ldots, t_s, f)$ ([H1], III, prop. 10.1(c)). □

There is a partial converse to this result, which will be very useful to prove that a rational curve is free or very free. The hypothesis that the tangent map to the evaluation map is surjective is weaker than the smoothness of that same map, and does not assume anything on the smoothness, or even reducedness, of the scheme $\mathrm{Mor}(\mathbf{P}^1, X; f|_B)$.

Proposition 4.9 *Let X be a smooth quasi-projective variety, let $f : \mathbf{P}^1 \to X$ be a rational curve, and let B be a finite subscheme of $\mathbf{P}^1$ of length b. If the tangent map to the evaluation map*

$$\mathrm{ev}:\ \begin{array}{ccc} (\mathbf{P}^1)^s \times \mathrm{Mor}(\mathbf{P}^1, X; f|_B) & \to & X^s \\ (t_1, \ldots, t_s, g) & \mapsto & (g(t_1), \ldots, g(t_s)) \end{array}$$

is surjective at some point of $(\mathbf{P}^1)^s \times \{f\}$, the rational curve f is $\min(2, b + s - 1)$-free.

PROOF. Upon replacing B by a subscheme and s by a smaller integer, we may assume $b + s \leq 3$. The hypothesis is that the map

$$\bigoplus_{i=1}^{s} T_{\mathbf{P}^1, t_i} \oplus H^0(\mathbf{P}^1, f^*T_X(-B)) \ \to\ \bigoplus_{i=1}^{s} T_{X, f(t_i)} \simeq \bigoplus_{i=1}^{s} (f^*T_X)_{t_i}$$

given by

$$(u_1, \ldots, u_s, \sigma) \mapsto \left(T_{t_1} f(u_1) + \sigma(t_1), \ldots, T_{t_s} f(u_s) + \sigma(t_s)\right)$$

is surjective for some $t_1, \ldots, t_s$. This implies that the evaluation

$$H^0(\mathbf{P}^1, f^*T_X(-B)) \to H^0(\mathbf{P}^1, f^*T_X) \to \bigoplus_{i=1}^{s} \left((f^*T_X)_{t_i} / \mathrm{Im}(T_{t_i} f)\right) \quad (4.3)$$

is surjective. We may assume that the dimension of X is at least 2, in which case this implies that no t_i is in B. There is a commutative diagram

$$\begin{array}{ccc} H^0(\mathbf{P}^1, f^*T_X(-B)) & \xrightarrow{\ e\ } & \bigoplus_{i=1}^{s}(f^*T_X)_{t_i} \\ \uparrow & & \uparrow {\scriptstyle (T_{t_1} f, \ldots, T_{t_s} f)} \\ H^0(\mathbf{P}^1, T_{\mathbf{P}^1}(-B)) & \xrightarrow{\ e'\ } & \bigoplus_{i=1}^{s} T_{\mathbf{P}^1, t_i} \end{array}$$

Since no t_i is in B and $\deg(T_{\mathbf{P}^1}(-B)) \geq s - 1$, the map e' is surjective; hence the image of e contains

$$\mathrm{Im}(T_{t_1} f, \ldots, T_{t_s} f)$$

Since the map (4.3) is surjective, e is surjective. With the notation (4.1), it follows that each map

$$H^0(\mathbf{P}^1, \mathscr{O}_{\mathbf{P}^1}(a_j - b)) \to H^0(\mathbf{P}^1, \mathscr{O}_{\mathbf{P}^1}(a_j)) \to \bigoplus_{i=1}^{s} k_{t_i}$$

for $j = 1, \ldots, n$, is surjective, hence $a_j - b \geq s - 1$ and f is $(b + s - 1)$-free. $\square$

4.10. *When the characteristic is zero, and there are*

- *a subscheme B of $\mathbf{P}^1$ of length b,*
- *a map $g : B \to X$,*
- *a subscheme M of $\mathrm{Mor}_{>0}(\mathbf{P}^1, X; g)$*

such that the evaluation map

$$\mathrm{ev}_M : (\mathbf{P}^1)^s \times M \to X^s$$

is dominant, the rational curves corresponding to points in some nonempty open subset of M are $\min(2, b + s - 1)$-free and pass through s general points of X.

Indeed, by [H1] (III, prop. 10.6), the tangent map to ev_M, hence also to ev, is surjective on some nonempty open subset of $\mathbf{P}^1 \times M$, and the assertion follows from Proposition 4.9.

We apply this principle in the proof of the next result.

Corollary 4.11 *Let X be a smooth quasi-projective variety.*

(a) *If X contains a free rational curve, X is uniruled.*

(b) *Conversely, if the characteristic is zero and X is uniruled and projective, there exists a free rational curve through a general point of X.*

In positive characteristic p, the conclusion of item (b) fails: we will see in Section 4.4 that the Fermat hypersurface of degree $p^r + 1$ in $\mathbf{P}^N$, with $N \geq 4$ and $r > 0$, is uniruled by lines, none of which are free. When $p^r > N$, this hypersurface has ample canonical divisor, hence contain no free rational curves. Also, the cyclic cover of degree p of $\mathbf{P}^n$, with $n \geq p$, branched along a general hypersurface of degree p is a Fano (hence uniruled) variety which has no free rational curves ([K1], th. V.5.11).

It is shown in Exercise 4.8.3 that a smooth projective uniruled variety X of dimension n defined over an algebraically closed field of characteristic zero is covered by rational curves $f : \mathbf{P}^1 \to X$ such that

$$f^* T_X \simeq \mathscr{O}_{\mathbf{P}^1}(2) \oplus \mathscr{O}_{\mathbf{P}^1}(1)^r \oplus \mathscr{O}_{\mathbf{P}^1}^{n-1-r}$$

where r is an integer that is equal to the dimension of the subvariety of X swept out by deformations of f passing through a fixed general point of X, minus 1. It is equal to $n - 1$ if and only if f is very free (compare with Proposition 4.20).

These curves are called *minimal*.

PROOF OF THE COROLLARY. If X contains a free rational curve $f : \mathbf{P}^1 \to X$, the evaluation map is smooth at $(0, [f])$. It follows that the restriction of the evaluation map to the unique component of $\mathrm{Mor}_{>0}(\mathbf{P}^1, X)$ that contains $[f]$ is dominant and X is uniruled.

Assume conversely that X is uniruled and projective. By Remark 4.2(3), the evaluation map is dominant on some irreducible component M of $\mathrm{Mor}_{>0}(\mathbf{P}^1, X)$; hence, when the characteristic is zero, the rational curve corresponding to a general point of M passes through a general point of X and is free by 4.10 (with $b = 0$ and $s = 1$).   □

Corollary 4.12 *Assume the characteristic is zero. If X is a smooth projective uniruled variety, $H^0(X, \mathcal{O}_X(mK_X))$ vanishes for all positive integers m.*

4.13. The converse is conjectured to hold, and it does in dimensions at most 3 (in characteristic zero). The references in the literature are as follows. By Mori's program, which is known to work in dimension 3, a nonuniruled variety is birational to a variety with nef canonical divisor, and the Kodaira dimension of such a variety is nonnegative (the case $H^1(X, \mathcal{O}_X) \neq 0$ is done in [U2] (th. 4.5), and the case $H^1(X, \mathcal{O}_X) = 0$ is Theorem 9.0.6 of Shepherd–Barron's article in [K8]; see also [MP], th. IV.3.1), which exactly means that one of the spaces $H^0(X, \mathcal{O}_X(mK_X))$ is nonzero. The conclusion follows, because the plurigenera $h^0(X, \mathcal{O}_X(mK_X))$ for $m > 0$ are birational invariants (see 7.1).

PROOF OF THE COROLLARY. By Corollary 4.11, there is a free rational curve $f : \mathbf{P}^1 \to X$ through a general point of X. Since f^*K_X has negative degree by Example 4.7(1), any section of $\mathcal{O}_X(mK_X)$ must vanish on $f(\mathbf{P}^1)$, hence on a dense subset of X, hence on X.   □

The next results says that a rational curve through a very general point (i.e., outside the union of a countable number of proper subvarieties) of a smooth variety is free (in characteristic zero).

Proposition 4.14 *Assume the characteristic is zero. Let X be a smooth quasi-projective variety. There exists a subset X^{free} of X which is the intersection of countably many dense open subsets of X, such that any rational curve on X whose image meets X^{free} is free.*

PROOF. The space $\mathrm{Mor}(\mathbf{P}^1, X)$ has at most countably many irreducible components, which we will denote by $(M_i)_{i \in \mathbf{N}}$. Let $\mathrm{ev}_i : \mathbf{P}^1 \times M_i \to X$ be the evaluation maps.

Denote by U_i a dense open subset of X such that the tangent map to ev_i is surjective at each point of $(\mathrm{ev}_i)^{-1}(U_i)$ (the existence of U_i follows from [H1], III, prop. 4.6 and uses the hypothesis that the characteristic is zero; if ev_i is not dominant, one may simply take for U_i the complement of the closure of the image of ev_i). We let X^{free} be the intersection $\bigcap_{i \in \mathbf{N}} U_i$.

Let $f : \mathbf{P}^1 \to X$ be a curve whose image meets X^{free}, and let M_i be an irreducible component of $\text{Mor}(\mathbf{P}^1, X)$ that contains $[f]$. By construction, the tangent map to ev_i is surjective at some point of $\mathbf{P}^1 \times \{[f]\}$, hence so is the tangent map to ev. It follows from Proposition 4.9 that f is free. $\square$

The proposition is interesting only when X is uniruled (otherwise, the set X^{free} is more or less the complement of the union of all rational curves on X). It is also useless when the ground field is countable, because X^{free} may be empty.

Examples 4.15 (1) If $\varepsilon : X \to \mathbf{P}^n$ is the blow-up of one point, X^{free} is the complement of the exceptional divisor E. Indeed, there is a $\mathbf{P}^1$-bundle $g : X \to \mathbf{P}^{n-1}$ of which E is a section, and an exact sequence

$$0 \to g^* \mathcal{O}_{\mathbf{P}^{n-1}}(1) \otimes \mathcal{O}_X(2E) \to T_X \xrightarrow{Tg} g^* T_{\mathbf{P}^{n-1}} \to 0$$

of vector bundles. If $f : \mathbf{P}^1 \to X$ is a rational curve not contained in E, the pull-back by f of this exact sequence expresses $f^* T_X$ as an extension of two nonnegative vector bundles, hence f is free.

Alternatively, in characteristic zero, one can use the fact that since the group G of automorphisms of X acts transitively on the complement of E, the G-equivariant map

$$\begin{array}{ccc} \mathbf{P}^1 \times G & \to & X \\ (t, \sigma) & \mapsto & \sigma(f(t)) \end{array}$$

is dominant, hence smooth, and f is free by Proposition 4.9.

(2) In characteristic zero, the proof above shows that if X is a smooth projective variety acted on by an affine algebraic group with a dense orbit, one may take X^{free} to be this dense orbit.

(3) On the blow-up X of $\mathbf{P}^2$ at nine general points, there are countably many rational curves with self-intersection -1 (see 6.6 and [H1], V, ex. 4.15(e)), which are not free by Example 4.7(3). When $\mathbf{k}$ is uncountable, X^{free} is not open.

The proposition will often be used together with the following remark. Let

$$\begin{array}{ccc} \mathscr{C} & \xrightarrow{F} & X \\ \downarrow{\scriptstyle \pi} & & \\ T & & \end{array}$$

be a flat family of curves on X parametrized by a variety T. If *the base field is uncountable* and one of these curves meets X^{free}, the same is true for a very general curve in the family.

Indeed, X^{free} is the intersection of a countable *nonincreasing* family $(U_i)_{i \in \mathbf{N}}$ of open subsets of X. Let $\mathscr{C}_t$ be the curve $\pi^{-1}(t)$. The curve $F(\mathscr{C}_t)$

meets X^{free} if and only if $\mathscr{C}_t$ meets $\bigcap_{i \in \mathbf{N}} F^{-1}(U_i)$. We have

$$\pi\left(\bigcap_{i \in \mathbf{N}} F^{-1}(U_i)\right) = \bigcap_{i \in \mathbf{N}} \pi(F^{-1}(U_i))$$

Let us prove it. The right-hand side contains the left-hand side. If t is in the right-hand side, the $\mathscr{C}_t \cap F^{-1}(U_i)$ form a nonincreasing family of nonempty open subsets of $\mathscr{C}_t$. Since the base field is uncountable, their intersection is nonempty. This means exactly that t is in the left-hand side.

Since π, being flat, is open ([Gr3], th. 2.4.6), this proves that the set of t such that $f_t(\mathbf{P}^1)$ meets X^{free} is the intersection of a countable family of open subsets of T.

This is expressed by the following principle:

4.16. A very general deformation of a curve that meets X^{free} has the same property.

4.3   Very Free Rational Curves and Rationally Connected Varieties

In Section 4.2, we studied the relationships between uniruledness and the existence of free rational curves on a smooth projective variety. We show here that there is an analogous relationship between rational connectedness and the existence of very free rational curves. For example, for a smooth projective variety defined over an algebraically closed field of characteristic zero, rational connectedness is equivalent to the existence of a single very free rational curve. This is the content of the following corollary.

Corollary 4.17 *Let X be a smooth projective variety.*

(a) *Assume X contains a very free rational curve. There is a very free rational curve through a general finite subset of X. In particular, X is rationally connected.*

(b) *Conversely, if the characteristic is zero and X is rationally connected, there exists a very free rational curve through a general point of X.*

The result will be strengthened in Corollary 4.28 where it is proved that in characteristic zero, there is on a smooth projective rationally connected variety a very free rational curve through *any* given finite subset (see also Exercise 4.8.6, where it is shown that tangents may also be prescribed).

PROOF OF THE COROLLARY. Assume that X contains a very free rational curve $f : \mathbf{P}^1 \to X$. By composing f with a finite map $\mathbf{P}^1 \to \mathbf{P}^1$ of degree r, we get an r-free curve. By Proposition 4.8 (applied with $B = \varnothing$), there is a deformation of this curve that passes through $r + 1$ general points of X. The rest of the proof is the same as in Corollary 4.11.   □

Corollary 4.18 *Assume the characteristic is zero. Let X be a smooth projective rationally connected variety.*

(a) $H^0(X, (\Omega_X^p)^{\otimes m})$ *vanishes for any positive integers m and p. In particular, $\chi(X, \mathcal{O}_X) = 1$.*

(b) *Any connected finite étale cover of X is trivial.*[3]

(c) *When $\mathbf{k} = \mathbf{C}$, the variety X is simply connected.*

Again, a converse to (a) is conjectured to hold: if $H^0(X, (\Omega_X^1)^{\otimes m})$ vanishes for all positive integers m, the variety X should be rationally connected. This is proved in dimension at most 3 in [KMM2], th. (3.2).

Singular rational varieties may be not simply connected (this is already the case for a nodal rational curve; see Exercise 4.8.2).

PROOF OF THE COROLLARY. By Corollary 4.17, there is a very free rational curve $f : \mathbf{P}^1 \to X$ through a general point of X. The vector bundle $f^*\Omega_X^1$ is a direct sum of line bundles of negative degree, hence any section of $(\Omega_X^p)^{\otimes m}$ must vanish on $f(\mathbf{P}^1)$, hence on a dense subset of X, hence on X. This proves (a).

By Hodge theory, $H^m(X, \mathcal{O}_X)$ vanishes for $m > 0$, hence $\chi(X, \mathcal{O}_X) = 1$. Let $\pi : \tilde{X} \to X$ be a connected finite étale cover; $\tilde{X}$ is rationally connected by Remark 4.4(5), hence $\chi(\tilde{X}, \mathcal{O}_{\tilde{X}}) = 1$. But $\chi(\tilde{X}, \mathcal{O}_{\tilde{X}}) = \deg(\pi)\chi(X, \mathcal{O}_X)$ hence π is an isomorphism. This proves (b).

Assume $\mathbf{k} = \mathbf{C}$. To prove (c), we follow an idea of [C2]. By Corollary 4.17, there exist a very free rational curve $f : \mathbf{P}^1 \to X$ and, by Proposition 4.8, a smooth open quasi-projective subset M of $\mathrm{Mor}(\mathbf{P}^1, X; 0 \mapsto f(0))$ over which the evaluation map $\mathrm{ev} : \mathbf{P}^1 \times M \to X$ is dominant.

Lemma 4.19 *Let X and Y be complex algebraic varieties, with Y normal, and let $f : X \to Y$ be a dominant morphism. The image of the induced morphism $\pi_1(f) : \pi_1(X) \to \pi_1(Y)$ has finite index.*

PROOF. The first remark is that if A is an irreducible analytic space and B a proper closed analytic subspace, $A - B$ is connected. The second remark is that the universal cover $\pi : \tilde{Y} \to Y$ is irreducible. Indeed, Y being normal is locally irreducible in the classical topology, hence so is $\tilde{Y}$. Since it is connected, it is irreducible.

Now if Z is a proper subvariety of Y, its inverse image $\pi^{-1}(Z)$ is a proper subvariety of $\tilde{Y}$; hence $\pi^{-1}(Y - Z)$ is connected by the two remarks above. This means exactly that the map $\pi_1(Y - Z) \to \pi_1(Y)$ is surjective. So we may replace Y with any dense open subset, and assume that Y is smooth.

We may also replace X with a desingularization and assume that it is smooth. Let $\overline{X}$ be a smooth compactification of X. We may replace X with

---

[3]The variety X is said to be *algebraically simply connected*.

the closure in $\overline{X} \times Y$ of the graph of f and assume that f is *proper*. Since the map $\pi(X) \to \pi_1(\overline{X})$ is surjective by the remark above, this does not change the cokernel of $\pi_1(f)$.

Finally, we may, by shrinking Y again, assume that f is smooth (this "generic smoothness" statement follows from Sard's theorem). The finite morphism in the Stein factorization of f is then étale. We may therefore assume that the fibers of f are connected. It is then classical that f is locally $\mathscr{C}^\infty$-trivial with compact fiber F, and the long exact homotopy sequence

$$\cdots \to \pi_1(F) \to \pi_1(X) \to \pi_1(Y) \to \pi_0(F) \to 0$$

of a fibration gives the result.   □

The composition of ev with the injection $\iota : \{0\} \times M \hookrightarrow X$ is constant, hence

$$\pi_1(\iota) \circ \pi_1(\text{ev}) = 0$$

Since $\mathbf{P}^1$ is simply connected, $\pi_1(\iota)$ is bijective, hence $\pi_1(\text{ev}) = 0$. Since ev is dominant, the lemma implies that the image of $\pi_1(\text{ev})$ has finite index. Therefore, the group $\pi_1(X)$ is finite, hence trivial by (b).   □

We finish this section with an analog of Proposition 4.14: on a *smooth* projective variety defined over an algebraically closed field *of characteristic zero*, a rational curve through a fixed point and a very general point is very free.

Proposition 4.20 *Assume the characteristic is zero. Let X be a smooth quasi-projective variety and let x be a point in X. There exists a subset X_x^{free} of X, which is the intersection of countably many dense open subsets of X, such that any rational curve on X passing through x and whose image meets X_x^{free} is very free.*

PROOF. The space $\text{Mor}(\mathbf{P}^1, X; 0 \mapsto x)$ has at most countably many irreducible components, which we will denote by $(M_i)_{i \in \mathbf{N}}$. Let $\text{ev}_i : \mathbf{P}^1 \times M_i \to X$ be the evaluation maps.

Denote by U_i a dense open subset of X such that the tangent map to ev_i is surjective at each point of $(\text{ev}_i)^{-1}(U_i)$, and let X_x^{free} be the intersection of the U_i. Let $f : \mathbf{P}^1 \to X$ be a curve with $f(0) = x$ whose image meets X^{free}, and let M_i be an irreducible component of $\text{Mor}(\mathbf{P}^1, X; 0 \mapsto x)$ that contains $[f]$. By construction, the tangent map to ev_i is surjective at some point of $\mathbf{P}^1 \times \{[f]\}$, hence so is the tangent map to ev. It follows from Proposition 4.9 that f is very free.   □

Again, this proposition is interesting only when X is rationally connected and the ground field is uncountable. The reader will have noticed that the proof is identical to that of Proposition 4.14, and that it shows that *given a subscheme B of X of length $b \le 2$, there exists a subset X_B^{free} of X, which is the intersection of countably many dense open subsets of X, such that any rational curve on X containing B and whose image meets X_B^{free} is b-free.*

4.4 Lines on a Complete Intersection Revisited

Let X be a subvariety of $\mathbf{P}^N$. We defined in Section 2.4 the scheme $F(X)$ of lines contained in X. It is the quotient of the scheme $\mathrm{Mor}_1(\mathbf{P}^1, X)$ by the action of the automorphism group of $\mathbf{P}^1$.

Assume X is a smooth complete intersection defined by equations of degrees $d_1, \ldots, d_s$ and set $|\mathbf{d}| = d_1 + \cdots + d_s$. A canonical divisor is $(|\mathbf{d}| - N - 1)H$, where H is a hyperplane section; X is therefore a Fano variety if and only if $|\mathbf{d}| \leq N$.

When $|\mathbf{d}| < N$, Proposition 2.13 says that the variety X is uniruled by lines. Moreover, when X is general, a general line on X is free (this follows also from (the proof of) Corollary 4.11 in characteristic zero). When $|\mathbf{d}| = N$, one can show that X is uniruled by conics.

For $|\mathbf{d}| > N$, the canonical divisor is nef so X contains no free rational curve. In characteristic zero, this implies that X is not uniruled (Corollary 4.11). However, we saw in 2.15 that when the characteristic p of $\mathbf{k}$ is positive, the Fermat hypersurface X of dimension at least 2 and degree $p^r + 1$ is uniruled by lines (it is even unirational by Exercise 2.5.1): the moduli space $\mathrm{Mor}_1(\mathbf{P}^1, X)$ is smooth, but the evaluation map

$$\mathrm{ev} : \mathbf{P}^1 \times \mathrm{Mor}_1(\mathbf{P}^1, X) \to X$$

is *not separable*.

4.5 Rationally Chain-Connected Varieties

We now study varieties for which two general points can be connected by a chain of rational curves (so this property is weaker than rational connectedness). For the same reasons as in Section 4.1, we have to modify slightly this geometric definition. We will eventually show (in Corollary 4.28) that rational chain-connectedness implies rational connectedness for *smooth* varieties in *characteristic zero*.

Recall that the ground field $\mathbf{k}$ is algebraically closed.

Definition 4.21 *A variety X is called* rationally chain-connected *if it is proper and there exist a variety T and a subscheme $\mathscr{C}$ of $T \times X$ such that:*

- *the fibers of the projection $\mathscr{C} \to T$ are (connected proper) curves with only rational components;*

- *the projection $\mathscr{C} \times_T \mathscr{C} \to X \times X$ is dominant.*

Remarks 4.22 (1) Rational chain-connectedness is *not* a birational property: the projective cone over an elliptic curve E is rationally chain-connected (pass through the vertex to connect any two points by a rational chain of length 2), but its canonical desingularization (a $\mathbf{P}^1$-bundle over E) is not. However, it is a birational property among *smooth* projective varieties in characteristic zero, because it is then equivalent to rational connectedness (Corollary 4.28).

(2) A general pair of points on a rationally chain-connected variety can be connected by a chain of rational curves. The converse is true when the base field is uncountable.[4] In fact, since a rationally chain-connected variety is proper, *any* two points can be connected by a chain of rational curves. This may be seen using a variant of the proof of Lemma 3.7 ([K1], cor. II.2.4).

(3) If X is a variety defined over an (algebraically closed) field k and K is an (algebraically closed) extension of k, the variety X_k is rationally chain-connected if and only if X_K is rationally chain-connected (proceed as in Remark 4.2(5), using (2)). This remark will allow us to assume when needed that the ground field is uncountable.

4.6   Smoothing Trees of Rational Curves

As promised, we are now on our way to prove that rational chain-connected-ness implies rational connectedness for *smooth* varieties in *characteristic zero* (this will be proved in Corollary 4.28). For that, we will need to smooth a chain of rational curves connecting two points, and this can be done when the links of the chain are free.

Definition 4.23 *A rational tree is a connected projective curve C whose singular points are nodes and which satisfies one of the equivalent following properties:*

(i) $\chi(C, \mathcal{O}_C) = 1$;

(ii) *the irreducible components of C are smooth rational curves and there are $\mathrm{Card}(\mathrm{Sing}(C)) + 1$ of them;*

(iii) *the irreducible components of C are smooth rational curves and they can be numbered as $C_1, \ldots, C_m$ in such a way that C_1 is any given component and, for each $i \in \{1, \ldots, m-1\}$, the curve C_{i+1} meets $C_1 \cup \cdots \cup C_i$ transversely in a single smooth point.*

Let us justify the equivalence of these properties. We begin with an elementary remark. If C is the union of curves $C_1', \ldots, C_r'$ with no common component and $\iota_i : C_i' \to C$ is the corresponding injection, there is an exact

---

[4]The proof follows the arguments used for uniruled or rationally connected varieties, using the closed subscheme $\mathrm{Rat}(X)$ of $\mathrm{Hilb}(X)$ that will be defined in 5.6, instead of $\mathrm{Mor}(\mathbf{P}^1, X)$.

sequence

$$0 \to \mathscr{O}_C \to \bigoplus_{i=1}^{r} \iota_{i*}\mathscr{O}_{C_i'} \to \bigoplus_{\substack{x \in C_i' \cap C_j' \\ 1 \leq i < j \leq r}} k_x \to 0$$

The associated long exact cohomology sequence yields

$$\chi(C, \mathscr{O}_C) = \sum_{i=1}^{r} \chi(C_i', \mathscr{O}_{C_i'}) - \sum_{1 \leq i < j \leq r} \mathrm{Card}(C_i' \cap C_j')$$

and

$$\sum_{i=1}^{r} h^1(C_i', \mathscr{O}_{C_i'}) \leq h^1(C, \mathscr{O}_C)$$

It is obvious that (iii) implies (ii). That (ii) implies (i) follows from the remark applied with $C_i' = C_i$. Finally, if (i) holds, we prove (iii) by induction on m. Let $C_2', \ldots, C_r'$ be the connected components of $\overline{C - C_1}$. By the remark above, we have $h^1(C_1, \mathscr{O}_{C_1}) = h^1(C_i', \mathscr{O}_{C_i'}) = 0$, hence C_1 is smooth rational and each C_i' satisfies (i), hence (iii), by induction. Moreover,

$$\sum_{i=2}^{r} \mathrm{Card}(C_1 \cap C_i') = r - 1$$

which means that each C_i' meets C_1 at a single point. We can number the components of C starting with C_1, then those of C_2' starting with the unique component that meets C_1, and so on. This proves (iii).

If C is a rational tree, there exist a smooth pointed 1-dimensional scheme (T, o) and a flat relative projective curve $\mathscr{C} \to T$ whose fiber over o is C, whereas all other fibers are smooth rational.[5] We say that $\mathscr{C} \to T$ is a *smoothing* of C. A morphism $f : C \to X$ is *smoothable* if there exist a smoothing $\mathscr{C} \to T$ and a morphism $\mathscr{C} \to X$ that coincides with f on C.

If $p_1, \ldots, p_r$ are distinct smooth points of C, we say that f is *smoothable keeping* $f(p_1), \ldots, f(p_r)$ *fixed* if there exist a smoothing $\mathscr{C} \to T$ with sections $\sigma_1, \ldots, \sigma_m$ such that σ_i passes through p_i, and a morphism $\mathscr{C} \to X$ that coincides with f on C and contracts each $\sigma_i(T)$ (to $f(p_i)$).

Proposition 4.24 *Let X be a smooth projective variety, let*

$$C = C_1 \cup \cdots \cup C_m$$

---

[5]Proceed by induction on m, starting from the trivial family $\mathbf{P}^1 \times \mathbf{P}^1 \to \mathbf{P}^1$ when $m = 1$. If $C = C_1 \cup \cdots \cup C_m$ is a rational tree, so is $C' = C_1 \cup \cdots \cup C_{m-1}$. Blowing up the surface $\mathscr{C}'$ given by the induction hypothesis at the point where C_m meets C' yields the surface $\mathscr{C}$.

be a rational tree,[6] and let $f : C \to X$ be a morphism. Let B be a set of r smooth points of C, with r_i points on C_i.

If the restriction of f to C_1 is $(r_1 - 1)$-free[7] and, for each $i \geq 2$, the restriction of f to C_i is r_i-free, f is smoothable into an $(r-1)$-free rational curve, keeping $f(B)$ fixed.

PROOF. Set $B = \{b_1, \ldots, b_r\}$. Let $\pi : \mathscr{C} \to (T, o)$ be a smoothing of C with sections $\sigma_1, \ldots, \sigma_r$ such that $\sigma_i(o) = b_i$. Denote by

$$g : \bigsqcup_{i=1}^{r} \sigma_i(T) \to X \times T$$

the morphism $\sigma_i(t) \mapsto (b_i, t)$. As mentioned in 2.10, T-morphisms from $\mathscr{C}$ to $X \times T$ extending g are parametrized by a T-scheme $\mathrm{Mor}_T(\mathscr{C}, X \times T; g)$ whose fiber at o is $\mathrm{Mor}(C, X; f|_B)$. Moreover, this scheme is smooth over T at $[f]$ when $H^1(C, f^*T_X(-B))$ vanishes (by the last two lines of 2.11).

We prove that this is the case by induction on m. The case $m = 1$ is obvious. Set

$$\begin{aligned}
C' &= C_1 \cup \cdots \cup C_{m-1} \\
B' &= B \cap C' \\
B_m &= B \cap C_m
\end{aligned}$$

There is an exact sequence

$$0 \to \mathscr{O}_{C_m}(-q - B_m) \to \mathscr{O}_C(-B) \to \mathscr{O}_{C'}(-B') \to 0$$

where q is the point of intersection of C' and C_m, which yields another exact sequence

$$H^1(C_m, f^*T_X(-q - B_m)) \to H^1(C, f^*T_X(-B)) \to H^1(C', f^*T_X(-B'))$$

By hypothesis and induction, both ends vanish. It follows that there exists a smooth 1-dimensional subvariety of $\mathrm{Mor}_T(\mathscr{C}, X \times T; g)$ passing through $[f]$ that dominates T. Let T' be its normalization. The morphism $T' \to \mathrm{Mor}_T(\mathscr{C}, X \times T; g)$ corresponds to a T'-morphism

$$\mathscr{C} \times_T T' \to X \times T'$$

which is the required smoothing of f to a rational curve $f_{t'} : \mathbf{P}^1 \to X$, keeping $f(B)$ fixed. Since the family $\mathscr{C} \times_T T' \to T'$ is flat,

$$H^1(\mathbf{P}^1, f_{t'}^*T_X(-\sigma_1(t') - \cdots - \sigma_r(t')))$$

---

[6]We assume that the components are numbered as in Definition 4.23(iii).
[7]See Definition 4.5.

vanishes by semicontinuity ([H1], III, th. 12.8) for t' general. It follows that $f_{t'}$ is $(r-1)$-free and proves the proposition.   □

4.25. A *rational comb* C is a rational tree C whose components can be denoted by $D, C_1, \ldots, C_m$ in such a way that each C_i meets the union of the other components in a single point q_i, which is on D. The C_i are the *teeth* of the comb, and D is the *handle*. A subcomb of C is a comb contained in C with the same handle.

We can construct a "universal" smoothing of the comb C as follows. Let $\mathscr{C}_m \to D \times \mathbf{A}^m$ be the blow-up of the (disjoint) union of the subvarieties $\{q_i\} \times \{y_i = 0\}$, where $y_1, \ldots, y_m$ are coordinates on $\mathbf{A}^m$. Fibers of $\pi : \mathscr{C}_m \to \mathbf{A}^m$ are subcombs of C, the number of teeth being the number of coordinates y_i that vanish at the point. Note that π is projective and flat, because its fibers are curves of the same genus 0 ([H1], III, th. 9.9; alternatively, one may use [H1], III, ex. 10.9). Let m' be a positive integer smaller than m, and embed $\mathbf{A}^{m'}$ in $\mathbf{A}^m$ as the subspace defined by the equations $y_i = 0$ for $m' < i \le m$. The inverse image $\pi^{-1}(\mathbf{A}^{m'})$ splits as the union of $\mathscr{C}_{m'}$ and $m - m'$ disjoint copies of $\mathbf{P}^1 \times \mathbf{A}^{m'}$.

The following smoothing result is harder: assuming only that the restriction of the morphism to the teeth is free, we still get a partial smoothing if there are enough teeth.

Proposition 4.26 *Let C be a rational comb with m teeth and let $p_1, \ldots, p_r$ be points on its handle D that are smooth on C. Let X be a smooth projective variety and let $f : C \to X$ be a morphism whose restriction to each tooth of C is free. Assume*

$$m > K_X \cdot f_* D + (r-1) \dim(X) + \dim_{[f|_D]} \mathrm{Mor}(\mathbf{P}^1, X; f|_{\{p_1, \ldots, p_r\}})$$

There exists a subcomb C' of C with at least one tooth such that $f|_{C'}$ is smoothable, keeping $f(p_1), \ldots, f(p_r)$ fixed.

PROOF. We keep the notation of 4.25, except that we set $\mathscr{C} = \mathscr{C}_m$. Let σ_i be the constant section of π equal to p_i, and let

$$g : \bigsqcup_{i=1}^{r} \sigma_i(\mathbf{A}^m) \to X \times \mathbf{A}^m$$

be the morphism $\sigma_i(y) \mapsto (f(p_i), y)$. Since π is projective and flat, there is an $\mathbf{A}^m$-scheme

$$\rho : \mathrm{Mor}_{\mathbf{A}^m}(\mathscr{C}, X \times \mathbf{A}^m; g) \to \mathbf{A}^m$$

as defined in 2.10. We will show that a neighborhood of $[f]$ is not contracted by ρ.

Since the fiber of ρ at 0 is $\mathrm{Mor}(C, X; f|_{\{p_1, \ldots, p_r\}})$ (2.10), it is enough to show

$$\dim_{[f]} \mathrm{Mor}(C, X; f|_{\{p_1, \ldots, p_r\}}) < \dim_{[f]} \mathrm{Mor}_{\mathbf{A}^m}(\mathscr{C}, X \times \mathbf{A}^m; g) \quad (4.4)$$

By the estimate (2.3), the right-hand side of (4.4) is at least

$$-K_X \cdot f_* C + (1 - r) \dim(X) + m$$

The fiber of the restriction

$$\operatorname{Mor}(C, X; f|_{\{p_1,\ldots,p_r\}}) \to \operatorname{Mor}(D, X; f|_{\{p_1,\ldots,p_r\}})$$

is $\prod_{i=1}^m \operatorname{Mor}(C_i, X; f|_{\{q_i\}})$, so the left-hand side of (4.4) is at most

$$\dim_{[f|_D]} \operatorname{Mor}(D, X; f|_{\{p_1,\ldots,p_r\}}) + \sum_{i=1}^m \dim_{[f]} \operatorname{Mor}(C_i, X; f|_{\{q_i\}})$$

$$= \dim_{[f|_D]} \operatorname{Mor}(D, X; f|_{\{p_1,\ldots,p_r\}}) + \sum_{i=1}^m (-K_X \cdot f_* C_i)$$

$$< m - K_X \cdot f_* C - (r - 1) \dim(X)$$

(We use the local description of $\operatorname{Mor}(C_i, X; f|_{\{q_i\}})$ given in 2.9, the fact that $f|_{C_i}$ being free, $H^1(C_i, f^* T_X(-q_i)|_{C_i})$ vanishes, and the hypothesis.)

Let T be the normalization of a 1-dimensional subvariety of $\operatorname{Mor}_{\mathbf{A}^m}(\mathscr{C}, X \times \mathbf{A}^m; g)$ passing through $[f]$ and not contracted by ρ. The morphism from T to $\operatorname{Mor}_{\mathbf{A}^m}(\mathscr{C}, X \times \mathbf{A}^m; g)$ corresponds to a T-morphism

$$\mathscr{C} \times_{\mathbf{A}^m} T \to X \times T$$

After renumbering the coordinates, we may assume that $\{m'+1,\ldots,m\}$ is the set of indices i such that y_i vanishes on the image of $T \to \mathbf{A}^m$, where m' is a *positive* integer. By 4.25, $\mathscr{C} \times_{\mathbf{A}^m} T$ splits as the union of $\mathscr{C}' = \mathscr{C}_{m'} \times_{\mathbf{A}^{m'}} T$, which is flat over T, and some other "constant" components $\mathbf{P}^1 \times T$. The general fiber of $\mathscr{C}' \to T$ is $\mathbf{P}^1$, its central fiber is the subcomb C' of C with teeth attached at the points q_i with $1 \le i \le m'$, and $f|_{C'}$ is smoothable keeping $f(p_1),\ldots,f(p_r)$ fixed.   □

4.7   Rational Chain-Connectedness Implies Rational Connectedness

When the characteristic is zero, we prove that a *smooth* rationally chain-connected variety is rationally connected (recall that this is false for singular varieties; see Remark 4.22(1)). We will need a more precise result that gives control on the degrees.

The basic idea of the proof is to use Proposition 4.24 to smooth a rational chain connecting two points. The problem is to make *each* link free. This is achieved by adding lots of free teeth to each link and by deforming the resulting comb into a free rational curve, keeping the two endpoints fixed, in order not to lose connectedness of the chain.

Theorem 4.27 *Assume the characteristic is zero. Let X be a smooth projective variety and let H be an ample divisor on X such that $H - K_X$ is nef. Assume there exists d in $\mathbf{N} \cup \{+\infty\}$ such that any two points of X can be connected by a rational chain of total H-degree at most d. Any two points of X can be connected by a single very free rational curve of H-degree at most $2d^2(\dim(X) + d + 2)^{d-1}$.*

PROOF. We may assume that the ground field is uncountable. Let x_1 and x_2 be points of X. There exists a rational chain connecting x_1 and x_2, which can be described as the union of rational curves $f_i : \mathbf{P}^1 \to C_i \subset X$, for $i = 1, \ldots, s$, with $f_1(0) = x_1$, $f_i(\infty) = f_{i+1}(0)$, $f_s(\infty) = x_2$.

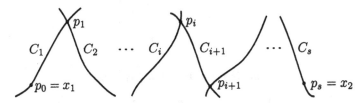

The rational chain connecting x_1 and x_2.

By Proposition 3.2, we may, if $\mathrm{Mor}(\mathbf{P}^1, X; f_i|_{\{0,\infty\}})$ has dimension at least 2 at $[f_i]$, deform f_i into a reducible curve with the same H-degree passing through the same endpoints. We will henceforth assume

$$\dim_{[f_i]} \mathrm{Mor}(\mathbf{P}^1, X; f_i|_{\{0,\infty\}}) \leq 1 \tag{4.5}$$

for $i = 1, \ldots, s$. Note that $s \leq d$ because H is ample.

Assume first that x_1 is in the subset X^{free} of X defined in Proposition 4.14, so that f_1 is free. We will construct by induction on i rational curves $g_i : \mathbf{P}^1 \to X$ with $g_i(0) = f_i(0)$ and $g_i(\infty) = f_i(\infty)$, whose image meets X^{free}.

When $i = 1$, take $g_1 = f_1$. Assume that g_i is constructed with the required properties. It is free, so the evaluation map

$$\begin{array}{rcl} \mathrm{ev}: \quad \mathrm{Mor}(\mathbf{P}^1, X) & \to & X \\ g & \mapsto & g(\infty) \end{array}$$

is smooth at $[g_i]$ (this is not exactly Proposition 4.8, but follows from its proof). Let T be an irreducible component of $\mathrm{ev}^{-1}(C_{i+1})$ that passes through $[g_i]$. It dominates C_{i+1}.

Apply Principle 4.16 to the family of rational curves on X parametrized by T: since one of the curves of this family, to wit g_i, meets X^{free}, so do very general members. Since they also meet C_{i+1} by construction, it follows that given a very general point x of C_{i+1}, there exists a deformation $h_x : \mathbf{P}^1 \to X$ of g_i which meets X^{free} and x.

Picking distinct very general points $x_1, \ldots, x_m$ in

$$C_{i+1} - \{f_{i+1}(0), f_{i+1}(\infty)\}$$

we get free rational curves $h_{x_1}, \ldots, h_{x_m}$ which, together with the handle C_{i+1}, form a rational comb with m teeth (as defined in 4.25) with a morphism to X whose restriction to the teeth is free.

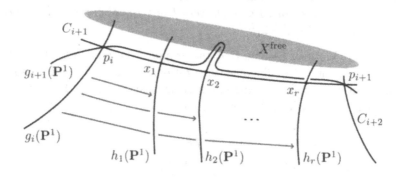

Replacing a link with a free link.

We want to apply Proposition 4.26 with the fixed points $f_{i+1}(0)$ and $f_{i+1}(\infty)$. Since we have, using (4.5),

$$
\begin{aligned}
& K_X \cdot C_{i+1} + \dim(X) + \dim_{[f_{i+1}]} \mathrm{Mor}(\mathbf{P}^1, X; f_{i+1}|_{\{0,\infty\}}) \\
\leq\ & H \cdot C_{i+1} + \dim(X) + 1 \qquad\qquad (4.6) \\
\leq\ & d + \dim(X) + 1
\end{aligned}
$$

we may take

- $m = d + \dim(X) + 2$ when d is finite;

- $m = H \cdot C_{i+1} + \dim(X) + 2$ if $d = +\infty$.

With these choices, a subcomb with at least one tooth can be smoothed. Since this subcomb meets X^{free}, so does a very general smooth deformation, by Principle 4.16 again.

So we managed to construct a rational curve $g_{i+1} : \mathbf{P}^1 \to X$ through $f_{i+1}(0)$ and $f_{i+1}(\infty)$, which meets X^{free}.

We will now bound the H-degree d_i of $g_i(\mathbf{P}^1)$ (when d is finite). The teeth of the comb were constructed as deformations of g_i, so have the same degree. It follows that we have the inequality

$$
d_{i+1} \leq m d_i + H \cdot C_{i+1}
$$

from which we get

$$
d_j \leq \sum_{i=1}^{j} m^{j-i} H \cdot C_i
$$

In the end, we get a chain of free rational curves connecting x_1 and x_2 whose total H-degree is at most

$$
\begin{aligned}
\sum_{j=1}^{s} d_j &\leq \sum_{1 \leq i \leq j \leq m} m^{j-i} H \cdot C_i \\
&\leq \frac{m^s - 1}{m - 1} \sum_{i=1}^{s} H \cdot C_i \qquad\qquad (4.7) \\
&\leq d \frac{m^s - 1}{m - 1} \\
&\leq d^2 m^{d-1}
\end{aligned}
$$

By Proposition 4.24, this chain can be smoothed leaving x_2 fixed. If M' is the subscheme of $\mathrm{Mor}(\mathbf{P}^1, X; 0 \mapsto x_2)$ that parametrizes morphisms with H-degree at most $d^2 m^{d-1}$, this means (as in the proof of Proposition 5.8) that x_1 is in the closure of the image of the evaluation map $\mathrm{ev} : \mathbf{P}^1 \times M' \to X$. Since x_1 is any point in X^{free}, and the latter is dense in X because the ground field is uncountable, ev is dominant. Let M be an irreducible component of M' such that $\mathrm{ev}|_{\mathbf{P}^1 \times M}$ is still dominant (when $d = +\infty$, we need again the fact that the ground field is uncountable). By 4.10, a general element of M corresponds to a very free rational curve through x_2 and a general point of X.

Since x_2 is *any* point of X, any two points of X can be connected by a chain of *two* very free rational curves (go through a general point of X), of total H-degree at most $2 d^2 m^{d-1}$.

By Proposition 4.24, this chain can be smoothed into a single very free rational curve, leaving the two given points fixed. $\qquad\square$

We saw in Corollary 4.17 that if there is a very free rational curve on a smooth projective variety, there is such a curve through a *general* finite subset. The next corollary shows that one can do better in characteristic zero.

Corollary 4.28 *Assume the characteristic is zero. Let X be a smooth projective variety. The following conditions are equivalent:*

(i) X *is rationally connected;*

(ii) X *is rationally chain-connected;*

(iii) *there is a very free rational curve through any finite subset of X.*

We will give in Proposition 5.8 a simpler proof of the equivalence of (i) and (ii) for varieties with Picard number 1. Also, we will see in Exercise 4.8.6 that on a smooth rationally connected projective variety in characteristic zero, we may find a very free curve passing through any finite subset of X with preassigned tangents.

PROOF OF THE COROLLARY. It is clear that (iii) implies (i), and that (i) implies (ii). Assume (ii) holds. We prove (iii) by induction on the number m of points.

The case $m = 2$ is covered by the theorem. Given points $x_1, \ldots, x_m$ on X, with $m \geq 3$, there is by induction a very free curve $f : \mathbf{P}^1 \to X$ passing through $x_1, \ldots, x_{m-1}$, and a very free curve passing through another point of $f(\mathbf{P}^1)$ and x_m. By composing f with a degree-$(m-2)$ cover $\mathbf{P}^1 \to \mathbf{P}^1$, we can make it $(m-2)$-free (see Definition 4.5), and, by Proposition 4.24, the resulting rational chain of length 2 can be smoothed into an $(m-2)$-free curve, keeping $x_1, \ldots, x_m$ fixed. $\square$

By composing it with a morphism $\mathbf{P}^1 \to \mathbf{P}^1$ of degree r, the rational curve in (iii) can be made r-free, with r greater than the number of points. By Exercise 4.8.5, a general deformation of that curve keeping the points fixed is an immersion if $\dim(X) \geq 2$ and an embedding if $\dim(X) \geq 3$. In other words, when $\dim(X) \geq 3$, given a finite set of points on the smooth rationally connected projective variety X, there is a smooth rational curve in X through these points with a normal bundle as ample as you want.

Corollary 4.29 *A smooth projective rationally chain-connected complex variety is simply connected.*

For example, a complex Fano variety is simply connected.

PROOF OF THE COROLLARY. A smooth projective rationally chain-connected variety is rationally connected by the corollary above, hence simply connected by Corollary 4.18. $\square$

Campana gives in [C4] a completely different proof of Corollary 4.29 which does not use the smoothing result. He shows more generally that the fundamental group of a compact Kähler manifold is finite if any two points can be connected by a chain of subvarieties whose fundamental group of the normalization has the same property. Campana later extended these results in [C5] to various classes of groups, such as groups with an abelian (resp. nilpotent, resp. solvable) subgroup of finite index.

4.8   Exercises

1. Show that the projective cone over a subvariety of a complex projective space is simply connected.

2. Describe the universal cover of a complex rational curve with one node and compute its fundamental group. A singular rational variety may not be simply connected.

3. Let X be a smooth projective uniruled variety of dimension n defined over an uncountable algebraically closed field of characteristic zero. Let d be the minimal degree of a free rational curve on X with respect to some fixed very ample line bundle.

   (a) Show that there exists a free rational curve $f : \mathbf{P}^1 \to X$ of degree d such that $x = f(0)$ is a point of X^{free}. Write

   $$f^*T_X \simeq \mathcal{O}_{\mathbf{P}^1}(a_1) \oplus \cdots \oplus \mathcal{O}_{\mathbf{P}^1}(a_n)$$

   with $a_1 \geq \cdots \geq a_n \geq 0$ as in (4.1).

   Let M be the component of $\mathrm{Mor}(\mathbf{P}^1, X)$ passing through $[f]$ and let, for each t in $\mathbf{P}^1$,

   $$\mathrm{ev}_t : \begin{array}{ccc} M & \to & X \\ g & \mapsto & g(t) \end{array}$$

   be the (dominant) evaluation map. Let M_x be an irreducible component of $\mathrm{ev}_0^{-1}(x)$.

   (b) Prove that $\mathrm{ev}_\infty(M_x)$, the subset of X swept out by the deformations of f passing through x, is a (closed) subvariety of X of dimension

   $$s = \dim(M_x) - 1 = \dim(M) - n - 1$$

   (c) Prove the equalities

   $$s = \mathrm{Card}\{i \mid a_i > 0\}$$

   and

   $$\dim(M) = \sum_{i=1}^n (a_i + 1)$$

   (d) Conclude

   $$f^*T_X \simeq \mathcal{O}_{\mathbf{P}^1}(2) \oplus \mathcal{O}_{\mathbf{P}^1}(1)^{s-1} \oplus \mathcal{O}_{\mathbf{P}^1}^{n-s}$$

   A rational curve $f : \mathbf{P}^1 \to X$ with this property is called *minimal*.

4. Let X be a smooth projective Fano variety. Using Exercise 4.8.3 above, show that there do not exist vector bundles E_1, E_2, and E_3 on X of rank at least 2 such that

   $$T_X \simeq E_1 \otimes E_2 \otimes E_3$$

   Find an example where T_X is the tensor product of two such vector bundles.

5. Let B be a finite subscheme of $\mathbf{P}^1$ with length b and let $f : \mathbf{P}^1 \to X$ be a free rational curve. With the notation (4.1), show the following.

(a) If $f|_B$ is unramified and $a_2 > b$, a general deformation of f keeping B fixed is unramified.

(b) If $f|_B$ is an embedding and $a_3 > b$, a general deformation of f keeping B fixed is an embedding.

6. Let X be a smooth projective rationally chain-connected variety defined over an algebraically closed field of characteristic zero. The aim of this exercise is to show that there exists on X a very free rational curve passing through any given finite subset $\{x_1, \ldots, x_b\}$ *with preassigned tangent directions* $\{\ell_1, \ldots, \ell_b\}$.

Let $\epsilon : \tilde{X} \to X$ be the blow-up of $\{x_1, \ldots, x_b\}$. We will identify $\ell_1, \ldots, \ell_b$ with points of $\tilde{X}$.

(a) Let $t_1, \ldots, t_b$ be distinct points on $\mathbf{P}^1$. Show that there exists a very free rational curve $\tilde{f} : \mathbf{P}^1 \to \tilde{X}$ such that $\tilde{f}(t_i) = \ell_i$ for each $i = 1, \ldots, b$.

(b) For each $i = 1, \ldots, b$, choose a nonzero tangent vector δ_i in $T_{\mathbf{P}^1, t_i}$. Compute the differential of the map

$$\mathrm{Mor}(\mathbf{P}^1, \tilde{X}; t_i \mapsto \ell_i) \quad \to \quad \bigoplus_{i=1}^{b} T_{\tilde{X}, \ell_i}$$

$$g \quad \mapsto \quad (T_{t_1} g(\delta_1), \ldots, T_{t_b} g(\delta_b))$$

(c) Conclude that there exists a very free rational curve $f : \mathbf{P}^1 \to X$ such that $f(t_i) = x_i$ and $\mathrm{Im}(T_{t_i} f) = \ell_i$ for each $i = 1, \ldots, b$.

Using Exercise 4.8.5(b), one may even construct a *smooth* rational curve with these properties if X has dimension at least 3.

5

The Rational Quotient

Let X be a variety. We define an equivalence relation $\mathscr{R}$ on X by saying that two points are $\mathscr{R}$-equivalent if they can be connected by a chain of rational curves (so that on a rationally chain-connected variety, two general points are $\mathscr{R}$-equivalent). The set of $\mathscr{R}$-equivalence classes is not in general an algebraic variety (there exist, for example, nonruled complex projective surfaces that contain countably many rational curves!). However, Campana realized in [C1] and [C4] that it is nevertheless possible to construct a very good substitute for the quotient if one throws away a countable union of proper subvarieties.

The first step is to construct a quotient for a family of rational curves parametrized by a quasi-projective scheme (e.g., the set of rational curves of fixed degree with respect to some projective embedding of X). It is in fact no more complicated to consider a family of subvarieties of X given by a diagram

$$\mathscr{C} \xrightarrow{\ F\ } X$$
$$\downarrow{\scriptstyle \pi}$$
$$T$$

and the associated equivalence relation: two points of X are $\mathscr{C}$-equivalent if they can be connected by a chain of subvarieties $F(\mathscr{C}_{t_1}) \cup \cdots \cup F(\mathscr{C}_{t_m})$. For a given point x of X, the sets $V_m(x)$ of points that can be connected to x by such a chain of fixed length m form an ascending chain of constructible subsets of X whose closures stabilize after at most n steps, where n is the dimension of X. For x general in X, the corresponding sets $\overline{V_n(x)}$ are either disjoint or equal, hence form the fibers of a rational map $\tau : X \dashrightarrow Y$ which is the required quotient for $\mathscr{C}$-equivalence. This process actually works only when F and π are *flat*. The construction of the quotient in this case is the object of Section 5.2, which comes after a Section 5.1 devoted to a few technical results having to do mainly with constructible sets, and which the reader is advised to skip in a first reading.

We present in Section 5.3 an application of the existence of a quotient in the flat case: in characteristic zero, a smooth projective uniruled variety with Picard number 1 is rationally connected. This is proved roughly as follows: on a variety with Picard number 1, any nonzero effective divisor is ample. This implies that the quotient $X \dashrightarrow Y$ for the flat family of *free* rational curves must be a point (because otherwise, the pull-back of a very ample divisor on Y would not be ample), hence a pair of general points of X are $\mathscr{R}$-equivalent. More precisely, they can be connected by a chain of free rational curves. This chain can be smoothed by the smoothing results of Section 4.6 into a single free rational curve connecting two general points of X.

The next step is to construct a quotient for $\mathscr{C}$-equivalence when F and π are *proper*. This is done in Section 5.4. The strategy is to restrict π over an open set where it is flat and use the existence of a quotient in the flat case to construct a partial quotient $\tau : X \to Y$. Using an induction, we prove that the rational quotient for the family

$$\mathscr{C} \xrightarrow{\ F\ } X \xrightarrow{\ \tau\ } Y$$
$$\downarrow{\scriptstyle \pi}$$
$$T$$

of subvarieties of Y yields the rational quotient for $\mathscr{C}$-equivalence on X. This brief description is oversimplified and there are unfortunately many technical details that render the actual proof rather painful. This is the most difficult part of this chapter.

To finish the construction of the rational quotient, we only need put together all the quotients $X \dashrightarrow Z_m$ associated with families of rational curves of degree bounded by m. This is not too difficult and done in Section 5.5. We obtain a rational map $\tau : X \dashrightarrow R(X)$ whose *very general* fibers are $\mathscr{R}$-equivalence classes.

It should be noted that Campana's construction in [C4] is different from the one presented here (which follows [KMM1] and [KMM3]) and works for not necessarily compact Kähler manifolds, as long as the map F in the above diagram is proper.

In Section 5.6, the existence of the rational quotient $\tau : X \dashrightarrow R(X)$ is used to prove that Fano varieties are rationally chain-connected: this is simply because we have proved in Theorem 3.22 that when X is a Fano variety, there exists a rational curve on X through a very general point of

X which is not contracted by the rational quotient map τ, *if $R(X)$ is not a point*. This cannot happen since very general fibers of τ are $\mathscr{R}$-equivalence classes; hence $R(X)$ is a point and X is rationally chain-connected. In characteristic zero, there is therefore by Corollary 4.28 a very free rational curve through any two points of X, and we even get an explicit bound on the degree of that curve that depends only on n.

This explicit bound yields, via an elementary argument, a (huge) bound on the self-intersection number $(-K_X)^n$ that depends only on n. We discuss in Section 5.8 an elementary differential-geometric derivation of such a bound for those complex Fano manifolds that have a positive Kähler–Einstein metric.

We explain in Section 5.9 (without giving any proofs) why the existence of such a bound implies that there are only finitely many deformation types of Fano varieties of fixed dimension. We discuss in Section 5.10 elementary examples of *singular* Fano surfaces to show that there are infinitely many deformation types when arbitrary singularities are allowed.

We conclude the chapter with a construction of Fano varieties with high degree $(-K_X)^n$. They are obtained as iterated projective bundles.

In this chapter, all schemes are (unless explicitly mentioned) of finite type over an algebraically closed field **k**.

5.1   Preliminaries

We gather in this section elementary results on the behavior of constructible sets under inverse images that we will need for the construction of the quotient, together with some facts about flat morphisms. Since nothing unexpected happens, the reader is advised to skip the whole section entirely, and come back to it when needed for later proofs.

Let X be a topological space. It is irreducible if it is not the union of two proper closed subsets.

Lemma 5.1 *Let V be a subset of a Noetherian topological space X. If*

$$\overline{V} = V_1 \cup \cdots \cup V_r$$

is an irredundant decomposition into irreducible components,

$$V = (V \cap V_1) \cup \cdots \cup (V \cap V_r)$$

is also an irredundant decomposition into irreducible components and

$$\overline{V \cap V_i} = V_i$$

for each i. In particular, V is irreducible if and only if $\overline{V}$ is.

PROOF. Assume first that $\overline{V}$ is irreducible, and let us prove that V is also irreducible. Write

$$V = (V \cap V_1) \cup (V \cap V_2)$$

with V_1 and V_2 closed in X. Then

$$\overline{V} = (\overline{V \cap V_1}) \cup (\overline{V \cap V_2})$$

hence we must have for instance $\overline{V} = \overline{V \cap V_1} \subset V_1$, which proves what we want.

Going back to the general case, we have

$$V = (V \cap V_1) \cup \cdots \cup (V \cap V_r)$$

hence

$$\overline{V} = (\overline{V \cap V_1}) \cup \cdots \cup (\overline{V \cap V_r})$$

and

$$V_1 \subset (\overline{V \cap V_1}) \cup V_2 \cup \cdots \cup V_r$$

Since V_1 is irreducible and not contained in $V_2, \ldots, V_r$, it must be contained in, hence equal to, $\overline{V \cap V_1}$. The intersection $V \cap V_1$ is then irreducible, since its closure is. Obviously, the $V \cap V_i$ are distinct since their closures are. This proves the lemma.   □

Lemma 5.2 *Let $\pi : X \to Y$ be a morphism and let V be an irreducible constructible subset of X. For y general in $\pi(V)$, we have*

$$\overline{V} \cap \pi^{-1}(y) = \overline{V \cap \pi^{-1}(y)}$$

PROOF. The right-hand side is contained in the left-hand side. Let us prove the other inclusion. By [H1], II, ex. 3.18(b), V contains a subset U that is dense open in $\overline{V}$. If $\overline{V} - U$ does not dominate $\pi(\overline{V})$, any y in $\pi(V)$ but not in $\pi(\overline{V} - U)$ satisfies

$$\overline{V} \cap \pi^{-1}(y) = U \cap \pi^{-1}(y) \subset \overline{V \cap \pi^{-1}(y)}$$

If $\overline{V} - U$ dominates $\pi(\overline{V})$, by a theorem of Chevalley ([H1], II, ex. 3.22(e)), the dimension at any point of a general fiber of $\overline{V} - U \to \pi(\overline{V})$ is

$$\dim(\overline{V} - U) - \dim(\pi(\overline{V}))$$

which is less than the dimension $\dim(\overline{V}) - \dim(\pi(\overline{V}))$ of the fiber of $\overline{V} \to \pi(\overline{V})$ at the same point. It follows that for y general in $\pi(V)$, the closed subset $(\overline{V} - U) \cap \pi^{-1}(y)$ of $\overline{V} \cap \pi^{-1}(y)$ is nowhere dense, hence that $U \cap \pi^{-1}(y)$ is dense in $\overline{V} \cap \pi^{-1}(y)$.   □

Lemma 5.3 *Let* $\pi : X \to Y$ *be a flat morphism and let W be a constructible subset of Y.*

(a) *One has* $\pi^{-1}(\overline{W}) = \overline{\pi^{-1}(W)}$.

(b) *Any irreducible component of $\pi^{-1}(W)$ dominates an irreducible component of W.*

(c) *If W is irreducible and π has irreducible fibers, $\pi^{-1}(W)$ is irreducible.*

PROOF. Let us first prove that any irreducible component of X dominates an irreducible component of Y. Let X' be an irreducible component of X and let Y' be an irreducible component of Y containing $\pi(X')$. The morphism $\pi^{-1}(Y') \to Y'$ is flat, so we may assume that Y is irreducible.

Let η be the generic point of X'. Then $\mathcal{O}_{X,\eta}$ is a flat $\mathcal{O}_{Y,\pi(\eta)}$-module hence has no torsion.[1] It follows that the map $\mathcal{O}_{Y,\pi(\eta)} \to \mathcal{O}_{X,\eta}$ is injective. Since its kernel defines the closure of $\pi(\eta)$ in Y, the latter is dense in Y.

Let us now prove (a). Assume there is an irreducible component V of $\pi^{-1}(\overline{W})$ not contained in $\overline{\pi^{-1}(W)}$. Then $V - \overline{\pi^{-1}(W)}$ is dense in V and is mapped to $\overline{W} - W$. It follows that $\pi(V)$ is contained in $\overline{W} - W$. Since W is constructible, it contains a subset U open and dense in $\overline{W}$. It follows that $\pi(V)$ is contained in $\overline{W} - U$, which is nowhere dense in $\overline{W}$. By what we have just seen, this contradicts the flatness of $\pi|_{\pi^{-1}(\overline{W})}$ and proves (a).

Since the morphism $\pi^{-1}(\overline{W}) \to \overline{W}$ is flat, any irreducible component of $\pi^{-1}(\overline{W}) = \overline{\pi^{-1}(W)}$ dominates an irreducible component of $\overline{W}$. By Lemma 5.1, any irreducible component of $\pi^{-1}(W)$ is dense in an irreducible component of $\pi^{-1}(\overline{W})$, hence dominates an irreducible component of $\overline{W}$, hence of W by the same lemma. This proves (b).

Finally, under the hypotheses of (c), assume that $\pi^{-1}(W)$ is the union of two closed subsets V_1 and V_2. Since the fibers of π are irreducible, W is the union of the constructible subsets $W_1 = \{x \in W \mid \pi^{-1}(x) \subset V_1\}$ and $W_2 = \{x \in W \mid \pi^{-1}(x) \subset V_2\}$. Since $\overline{W}$ is irreducible (Lemma 5.1), one of them, say W_1, must be dense in W. Since π is flat, item (a) then implies

$$\pi^{-1}(W) \subset \pi^{-1}(\overline{W}_1) = \overline{\pi^{-1}(W_1)} \subset V_1$$

hence $\pi^{-1}(W) = V_1$ since V_1 is closed in $\pi^{-1}(W)$. This proves that $\pi^{-1}(W)$ is irreducible. $\qquad\square$

5.4. We will need the following result ([Gr3], th. 6.9.1), to which we will refer as "generic flatness:" *let $\pi : X \to Y$ be a morphism. If Y is reduced, there exists a dense open subset U of Y such that the morphism $\pi^{-1}(U) \to U$ induced by π is flat.*

---

[1] If A is a domain and M is a flat A-module, for any nonzero x in A, the multiplication by x in A is injective, hence so is the multiplication by x in M by flatness.

5.5. Finally, we will also need the following result: *let $\pi : X \to Y$ be a morphism whose closed fibers are connected. If X is normal and Y irreducible, a general fiber of π is irreducible.*

This can be seen as follows. The set of points of Y whose fiber is geometrically connected is constructible ([Gr4], th. 9.7.7). Since it contains all closed points, the generic fiber F of π is geometrically connected, and normal (its local rings are local rings of X). Let K be the field of rational functions of Y, let K^s be a separable closure of K and let K^a be an algebraic closure of K^s.

The scheme $F^s = F \otimes_K K^s$ is connected, normal by [Gr3], prop. 6.7.4, hence irreducible. The map $F^a = F \otimes_K K^a \to F^s$ is a homeomorphism, hence F^a is also irreducible. But again, the set of points of Y whose fiber is geometrically irreducible is constructible ([Gr4], th. 9.7.7). Since it contains the generic point of Y, it is dense.

5.6. We will use sporadically the existence of the Hilbert scheme of a quasi-projective variety X. This locally Noetherian scheme $\mathrm{Hilb}(X)$ (constructed in [Gr5]) parametrizes[2] flat families of subschemes of X. Logically speaking, this paragraph should have come before Section 2.2, since the space $\mathrm{Mor}(Y, X)$ is constructed as an open subscheme of $\mathrm{Hilb}(Y \times X)$ (via the map that associates its graph to a morphism). However, since it will play a much less central role for us, we will only mention a few relevant facts.

Assume X is projective and let H be an ample divisor on X. If one fixes the Hilbert polynomial $P(m) = \chi(Z, \mathcal{O}_Z(mH))$ of the subscheme Z, the corresponding subscheme of $\mathrm{Hilb}(X)$ is projective ([Gr5], th. 3.2). For 1-dimensional subschemes, this amount to fixing the H-degree and the Euler characteristic.

Let $\mathrm{Rat}(X)$ be the union of the components of $\mathrm{Hilb}(X)$ whose general points correspond to (reduced connected) curves with rational components. If the degree of such a curve is bounded by d, so is the number of components, as well as the multiplicity of the singular points; hence the Euler characteristic is also bounded and the corresponding subscheme $\mathrm{Rat}_d(X)$ of $\mathrm{Hilb}(X)$ is *projective*. The same proof as in Lemma 3.7 shows that the irreducible components of the subscheme of X corresponding to any point of $\mathrm{Rat}(X)$ are all rational curves.

5.2   Quotient by a Flat Algebraic Relation

The construction of the rational quotient of a variety X can be done in the following more general setting.

---

[2]With the same meaning as in Section 2.2.

Given reduced quasi-projective schemes T and $\mathscr{C}$, possibly reducible, we start from what we will call a *basic diagram:*

$$\mathscr{C} \xrightarrow{\ \ F\ \ } X$$
$$\downarrow{\scriptstyle \pi}$$
$$T$$

(5.1)

For each point t of T, let $\mathscr{C}_t$ be the fiber $\pi^{-1}(t)$. We say that two points x and x' of X can be connected by a $\mathscr{C}$-chain of length m if there exist points $t_1, \ldots, t_m$ of T such that some connected component of $F(\mathscr{C}_{t_1} \cup \cdots \cup \mathscr{C}_{t_m})$ contains x and x'. When the fibers of π are connected (which will always be the case in practice, but this degree of generality is needed for the proof), this is the same as asking that there exist $t_1, \ldots, t_m$ in T such that $x \in F(\mathscr{C}_{t_1})$, each $F(\mathscr{C}_{t_i})$ meets $F(\mathscr{C}_{t_{i+1}})$ for $i = 1, \ldots, m-1$, and $x' \in F(\mathscr{C}_{t_m})$.

Two points of X are $\mathscr{C}$-equivalent if they can be connected by a $\mathscr{C}$-chain of finite length (we agree that any point can be connected to itself by a $\mathscr{C}$-chain of length 0).

Proposition 5.7 *In the diagram (5.1), assume that π is flat with irreducible fibers and that F is flat. There exists a dense open subset X^0 of X, a variety Y^0 and a morphism $\tau : X^0 \to Y^0$ such that[3]*

(a) *two points of X^0 have the same image by τ if and only if the closures in X of their $\mathscr{C}$-equivalence classes are the same;*

(b) *if x is in X^0, a general point of its fiber $\tau^{-1}(\tau(x))$ can be connected to x by a $\mathscr{C}$-chain of length $\dim(X) - \dim(Y^0)$.*

PROOF. By shrinking X (and $\mathscr{C}$), we may assume that X is quasi-projective and normal.

For each point x of X, let $V_m(x)$ be the set of points of X that can be connected to x by a $\mathscr{C}$-chain of length m. Let

$$\delta(x) = \max_{m \geq 1} \dim(\overline{V_m(x)})$$

First step: $\overline{V_{m+1}(x)} = \overline{V_m(x)}$ *for $m \geq \delta(x)$.*
Since

$$\pi(F^{-1}(x)) = \{t \in T \mid x \in \mathscr{C}_t\}$$

a point x' can be connected to x by a $\mathscr{C}$-chain of length m if and only if $\pi(F^{-1}(x))$ meets $\pi(F^{-1}(x'))$. It follows that the subsets $V_m(x)$ can be defined recursively by the formula

$$V_{m+1}(x) = F(\pi^{-1}(\pi(F^{-1}(V_m(x)))))$$

---

[3]Note that τ might not be a morphism of k-schemes in positive characteristic.

In particular, they are *constructible*.

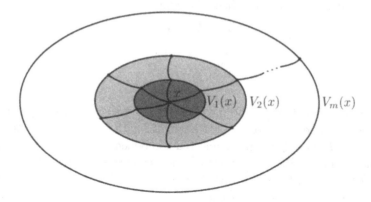

The constructible sets $V_m(x)$.

Let $(V_m^i)_i$ be the irreducible components of $V_m(x)$ and let $(W_m^{i,j})_j$ be those of $F^{-1}(V_m^i)$. Since F is flat, we have by Lemma 5.3

$$F^{-1}(\overline{V_m^i}) = \overline{F^{-1}(V_m^i)} = \bigcup_j \overline{W_m^{i,j}} \quad \text{and} \quad F(\overline{W_m^{i,j}}) = \overline{V_m^i}$$

Since π is flat with irreducible fibers, $\tilde{W}_m^{i,j} = \pi^{-1}(\pi(W_m^{i,j}))$ is irreducible by Lemma 5.3(c) and

$$\overline{V_{m+1}(x)} = \bigcup_{i,j} \overline{F(\tilde{W}_m^{i,j})}$$

where each $\overline{F(\tilde{W}_m^{i,j})}$ contains $\overline{V_m^i}$.

Call $\overline{V_m^i}$ *stable* if $\overline{F(\tilde{W}_m^{i,j})} = \overline{V_m^i}$ for all j, *unstable* otherwise. Note the following:

- if $\overline{V_m^i}$ is unstable, it is not an irreducible component of $\overline{V_{m+1}(x)}$;

- if $\overline{V_m^i}$ is stable and an irreducible component of $\overline{V_{m+1}(x)}$, it is stable as a component of $\overline{V_{m+1}(x)}$.

Let $V = \overline{F(\tilde{W}_m^{i,j})}$ be an unstable component of $\overline{V_{m+1}(x)}$. If the corresponding $\overline{V_m^i}$ is stable, it is equal to V by definition, which contradicts the second item. Hence $\overline{V_m^i}$ is unstable, and cannot be equal to V by the first item.

It follows that if $\overline{V_{m+1}(x)}$ has an unstable component V, then $\overline{V_m(x)}$ also has an unstable component of smaller dimension. In particular, the dimension of V is at least $m + 1$. The first step follows.

Second step: let n be the dimension of X. There exists an open dense subset X^0 of X such that, for any points x, x' in X^0,

$$x' \in \overline{V_n(x)} \iff \overline{V_n(x)} = \overline{V_n(x')}$$

The subset

$$V = \bigcup_{x \in X} \left(\{x\} \times V_n(x) \right)$$

of $X \times X$ is constructible. Indeed, if we define constructible subsets V_i of $X \times X$ inductively by taking for V_0 the diagonal in $X \times X$ and setting

$$V_{m+1} = (F \times \mathrm{Id}_X)((\pi \times \mathrm{Id}_X)^{-1}((\pi \times \mathrm{Id}_X)((F \times \mathrm{Id}_X)^{-1}(V_m))))$$

then $V = V_n$.

Let $q : V \to X$ the first projection. In positive characteristic, general fibers of q might not be reduced, but this can be fixed as follows: there exists a composition $F_r : X_r \to X$ of r Frobenius morphisms such that the morphism $q_r : (\overline{V} \times_X X_r)_{\mathrm{red}} \to X_r$ obtained from q by base change has reduced general fibers. This can be seen as follows: let η be the generic point of X and let $\overline{V}_\eta$ be the generic fiber of q. By [Gr3], prop. 4.6.6, there exists a purely inseparable finite extension K of $k(\eta)$ such that $(\overline{V}_\eta \times_{k(\eta)} K)_{\mathrm{red}}$ is geometrically reduced. Let p be the characteristic exponent of $\mathbf{k}$ (i.e., the characteristic if it is positive, 1 otherwise). There exists an integer r such that K is contained in $k(\eta)^{p^{-r}}$ (the field of p^rth roots of elements of $k(\eta)$ is some fixed algebraic closure), and by [Gr3], prop. 4.6.5, the scheme $(\overline{V}_\eta \times_{k(\eta)} k(\eta)^{p^{-r}})_{\mathrm{red}}$ is again geometrically reduced. This scheme is the generic fiber of q_r, hence general fibers of q_r are reduced by [Gr4], th. 9.7.7. As abstract schemes, X_r and X are isomorphic, F_r is the identity on points, and fibers of q and q_r are the same *as sets*.

There exists a dense open subset X_r^0 of X_r, with image X^0 in X, such that

- q is flat over X^0 (5.4) and q_r is flat over X_r^0 with reduced fibers ([Gr4], cor. 12.1.7(vi));

- $V_n(x)$ is dense in $q^{-1}(x)$ for all x in X^0 (Lemma 5.2).

Let x be a point of X^0. If $x' \in V_n(x)$, any point in $V_n(x')$ can be connected by a $\mathscr{C}$-chain of length n to x', which can itself be connected to x by a $\mathscr{C}$-chain of length n. It follows that $V_n(x')$ is contained in $V_{2n}(x)$, whose closure is by the first step equal to $\overline{V_n(x)}$. It follows that $\overline{V_n(x')}$ is contained in $\overline{V_n(x)}$, and there is equality by symmetry if x' is also in X^0.

By the second item, this means exactly that the second projection of $q^{-1}(V_n(x) \cap X^0)$ is $\overline{V_n(x)}$. Since q is flat over X^0, the set $q^{-1}(V_n(x) \cap X^0)$

is *dense* in $q^{-1}(\overline{V_n(x)} \cap X^0)$ (Lemma 5.3(a)); hence the second projection of $q^{-1}(\overline{V_n(x)} \cap X^0)$ is also $\overline{V_n(x)}$. All this implies

$$\overline{V_n(x')} = \overline{V_n(x)}$$

for all x' in $\overline{V_n(x)} \cap X^0$ and proves the second step.

Third step: construction of the quotient morphism τ.

The morphism q_r is flat over X_r^0 hence defines a morphism $\tau : X_r^0 \to \mathrm{Hilb}(X)$ which sends a point x to the point of the Hilbert scheme corresponding to the (reduced) subscheme $q_r^{-1}(x) = \overline{V_n(x)}$ of X.[4] The image of τ is constructible, hence contains an open subset Y^0 dense in its closure. Replace X_r^0 with the smaller open subset $\tau^{-1}(Y^0)$. Points x and x' of X_r^0 are in the same fiber of τ if and only if the *subschemes* $q_r^{-1}(x)$ and $q_r^{-1}(x')$ of X are equal. Since they are reduced, this is equivalent to $\overline{V_n(x)} = \overline{V_n(x')}$, which by the second step happens exactly when $x' \in \overline{V_n(x)}$. This proves (a).

Since q is flat, its fibers have constant dimension $\delta = n - \dim(Y^0)$ ([H1], III, prop. 9.5). By the first step, we have $\overline{V_\delta(x)} = \overline{V_n(x)}$ for all x in X^0. It follows that a general point in $\tau^{-1}(\tau(x))$ is in $V_\delta(x)$, hence can be connected to x by a $\mathscr{C}$-chain of length δ. This proves (b). Moreover, *any* two points x and x' in the same fiber of τ can be connected by a $\mathscr{C}$-chain of length 2δ, since they can be both connected to a general point of their fiber by a $\mathscr{C}$-chain of length δ. $\square$

5.3   Uniruled Varieties with Picard Number 1

We will take a break in order to present an application of the existence of a quotient for the equivalence relation associated with a flat family of curves to varieties with Picard number 1.[5] Historically (although this is not so old), this was the first case that was considered, because the proofs are easier (we will not need the difficult smoothing result of Proposition 4.26). Also, the results and the various bounds obtained are much better than in the general case.

Proposition 5.8 *If a smooth projective variety with Picard number 1 has a free rational curve, it is rationally connected.*

---

[4]See 5.6 for a few facts about the Hilbert scheme. Note that we are here identifying a point of X_r with its image on X.

[5]Recall that the Picard number of a proper variety X is the dimension of the vector space $N^1(X)_{\mathbf{R}}$.

More precisely, if such a variety X has a free rational curve of degree d with respect to some ample divisor, there is a free rational curve of degree at most $d \dim(X)$ through a general[6] pair of points of X.

Assume that the characteristic is zero. If X is uniruled with Picard number 1, it has a free rational curve (Corollary 4.11). The proposition applies and produces a rational curve through a general pair of points of X which is *very free* by 4.10. This is the case for example when X is a Fano variety, with $d = \dim(X) + 1$, where the degree is taken with respect to the anticanonical divisor (see Theorem 3.4, and compare with Proposition 5.16).

PROOF OF THE PROPOSITION. Free deformations of the given free rational curve are parametrized by a smooth quasi-projective variety M and the evaluation map $\mathrm{ev} : \mathbf{P}^1 \times M \to X$ is smooth (Proposition 4.8). Proposition 5.7 yields a quotient $\tau : X^0 \to Y^0$. If Y is a compactification of Y^0, the corresponding rational map $X \dashrightarrow Y$ is defined on an open set X^1 whose complement has codimension at least 2 (1.39). Let $\tau^1 : X^1 \to Y$ be the resulting morphism. Since $X - X^1$ has codimension at least 2, so does its inverse image by the smooth morphism ev in $\mathbf{P}^1 \times M$, hence it cannot dominate M. Therefore, there exists a dense open subset M^1 of M such that $\mathrm{ev}(\mathbf{P}^1 \times M^1)$ is contained in X^1.

By property (a) of Proposition 5.7,

$$\tau^1 \circ \mathrm{ev}^1 : \mathbf{P}^1 \times M^1 \to Y$$

contracts those $\mathbf{P}^1 \times \{t\}$ whose image by ev meets X^0. Since being contracted is a numerical property, it contracts $\mathbf{P}^1 \times \{t\}$ for *every* t in M^1.

Assume that Y is not a point. Let H^1 be the inverse image on X^1 of an ample effective divisor H_Y on Y and let H be its closure in X. Since $N^1(X)_{\mathbf{R}}$ has dimension 1, *any nonzero effective divisor on X is ample*. Let $t \in M^1$. The curve $\mathrm{ev}^1(\mathbf{P}^1 \times \{t\})$ meets H hence also H^1 since it is contained in X^1. But it is also contracted by τ^1 and, since H^1 is the pullback by τ^1 of the divisor H_Y on Y, its image by τ^1 is a point of H_Y. This implies that the image of $\tau^1 \circ \mathrm{ev}^1$ is contained in H_Y, which is absurd since this map is smooth hence dominant.

Consequently Y is a point and it follows from property (b) of Proposition 5.7 that a general pair of points of X can be connected by a chain of free rational curves of total degree at most $d \dim(X)$. We will now smooth this chain.

Let $e : M \to X \times X$ be the morphism $[f] \mapsto (f(0), f(\infty))$. We have just proved that there exists a dense open subset U of $X \times X$ such that, for any (x_1, x_2) in U, there is a chain $f : C \to X$ of *free* rational curves connecting

---

[6]It follows from the properness of the scheme $\mathrm{Rat}_{d \dim(X)}(X)$ discussed in 5.6 that *any* two points of X can be connected by a chain of rational curves of total degree at most $d \dim(X)$, but the links might not be free anymore.

x_1 and x_2, of total degree at most $d \dim(X)$. Let p_1 and p_2 be points of C such that $f(p_i) = x_i$. We may assume that p_1 and p_2 are smooth on C by throwing away extra components.

By Proposition 4.24, there exist a smoothing $\pi : \mathscr{C} \to (T, o)$ of C with a section σ_1 passing through p_1 and a morphism $F : \mathscr{C} \to X$ extending f and keeping x_1 fixed. Since π is projective, there exists a hyperplane section T' of $\mathscr{C}$ through p_2. Replacing π by its pull-back to the normalization of T', we may assume that there is also a section σ_2 of π such that $\sigma_2(o) = p_2$. By construction, $F(\sigma_1(t), \sigma_2(t))$ is in $e(M)$ for all $t \neq o$, hence (x_1, x_2) is in $e(M)$. It follows that e is dominant. Its image being constructible contains a dense open subset of $X \times X$, and this proves the proposition. $\square$

5.4   Quotient by a Proper Algebraic Relation

Starting again from a basic diagram (see (5.1))

$$
\begin{array}{ccc}
\mathscr{C} & \xrightarrow{\;F\;} & X \\
\downarrow{\scriptstyle \pi} & & \\
T & &
\end{array}
$$

where we assume that π and F are *proper*, we construct a *proper* quotient map, most of whose fibers are equivalence classes.[7] The chains however get longer than in the flat case. Part (a) of our next result is due to Campana. It was first proved in the analytic setting in [C1]. The estimation of (b) was later proved by Kollár, Miyaoka and Mori in [KMM3].

Theorem 5.9 *In the diagram (5.1), assume X is a proper normal variety and that F and π are proper. There exists a dense open subset X^* of X, a variety Y^*, and a proper morphism $\rho : X^* \to Y^*$ such that*

  (a) *each fiber of ρ is a $\mathscr{C}$-equivalence class;*

---

[7]The following result is Theorem IV.4.16 of [K1]. However, the proof presented on p. 220 of [K1] is incomplete (and so is the proof of Lemma 2 in [D]!). The reason is that one has to take into account that when reducing T to make π flat, for example, $\mathscr{C}$-equivalence classes may change. An example is provided by the basic diagram

$$
\begin{array}{ccc}
\tilde{\mathbf{P}}^2 & \xrightarrow{\;F\;} & \mathbf{P}^1 \\
\downarrow{\scriptstyle \pi} & & \\
\mathbf{P}^2 & &
\end{array}
$$

where π is the blow-up of a point. The quotient is a point (F maps the exceptional divisor onto $\mathbf{P}^1$), but by restricting to an open set where π is flat, the equivalence classes become points, and the quotient is the wrong one!

(b) *any two points of a fiber of ρ can be connected by a $\mathscr{C}$-chain of length* $2^{\dim(X)-\dim(Y^*)} - 1$.

Note that by 5.5, general fibers of ρ, hence also general $\mathscr{C}$-equivalence classes, are irreducible and proper. If X is smooth and the characteristic is zero, these classes are moreover smooth by generic smoothness.

PROOF OF THE THEOREM. *First reduction.* We may assume that F is surjective (otherwise, take $X^* = Y^* = X - F(\mathscr{C})$).

Second reduction. Let $\tilde{\mathscr{C}}$ be the normalization of $\mathscr{C}$ and let $\tilde{\mathscr{C}} \xrightarrow{\tilde{\pi}} \tilde{T} \xrightarrow{p} T$ be the Stein factorization of $\tilde{\mathscr{C}} \to \mathscr{C} \xrightarrow{\pi} T$. Assume we can construct a quotient $X^* \to Y^*$ for $\tilde{\mathscr{C}}$-equivalence as in the theorem. Any two points of X that can be connected by a $\tilde{\mathscr{C}}$-chain can be connected by a $\mathscr{C}$-chain of equal length. Conversely, any two points of X that can be connected by a $\mathscr{C}$-chain of length ℓ can be connected by a $\tilde{\mathscr{C}}$-chain of length $\ell \deg(p)$. So $\mathscr{C}$- and $\tilde{\mathscr{C}}$-equivalence classes are the same, and we can take the same quotient for $\mathscr{C}$-equivalence. So by performing this construction, *we may assume that $\mathscr{C}$ is normal and that π has connected fibers.*

Third reduction. Let $u : X' \to X$ be a proper morphism such that $u_* \mathcal{O}_{X'} \simeq \mathcal{O}_X$ (e.g., a proper birational morphism). Assume we can find a quotient $\rho' : X'^* \to Y^*$ as in the theorem for the basic diagram

$$\mathscr{C}' = \mathscr{C} \times_X X' \xrightarrow{F'} X'$$
$$\downarrow{\scriptstyle \pi'}$$
$$T$$

For any points $t_1, \ldots, t_m$ of T, we have

$$F'(\mathscr{C}'_{t_1}) \cup \cdots \cup F'(\mathscr{C}'_{t_m}) = u^{-1}(F(\mathscr{C}_{t_1}) \cup \cdots \cup F(\mathscr{C}_{t_m}))$$

Since the fibers of u are connected ([H1], III, cor. 11.3), it follows that two points x'_1 and x'_2 of X' can be connected by a $\mathscr{C}'$-chain of length m if and only if the points $u(x'_1)$ and $u(x'_2)$ of X can be connected by a $\mathscr{C}$-chain of length m (when $m = 0$, we need the surjectivity of F). Fibers of u are therefore contained in $\mathscr{C}'$-equivalence classes. It follows that

- $X'^* = u^{-1}(u(X'^*))$, hence

$$X^* = u(X'^*) = X - u(X' - X'^*)$$

  is open in X since u is proper;

- ρ' contracts fibers of u hence factors as $X'^* \to X^* \xrightarrow{\rho} Y$, where ρ is proper (Lemma 1.15).

Moreover, the map ρ satisfies properties (a) and (b) of the theorem.

Let us now begin the proof. By the third reduction and Chow's lemma (see Remark 4.2(4)), we may assume that X is projective. By the second reduction, we may assume that $\mathscr{C}$ is normal and π has connected fibers.

Let U be a dense open subset of X over which F is flat (5.4). Since $F^{-1}(U)$ is normal (and nonempty), there exists a subset T^1 of $\pi(F^{-1}(U))$ open in $\pi(F^{-1}(U))$ over which $\pi|_{F^{-1}(U)}$ is flat with irreducible fibers (5.4 and 5.5). Let $\mathscr{C}^1$ be the nonempty open subset $\pi^{-1}(T^1)$ of $\mathscr{C}$.

By Proposition 5.7 applied to the basic diagram

$$
\begin{array}{ccc}
\mathscr{C}^1 & \xrightarrow{\ F\ } & X \\
\downarrow{\scriptstyle \pi} & & \\
T^1 & &
\end{array}
$$

there exists an open subset X^0 of U and a quotient $\tau : X^0 \to Y^0$ with the required properties for $\mathscr{C}^1$-equivalence (where τ is not necessarily a morphism of k-schemes; see footnote 3, p. 117). Since any $\mathscr{C}^1$-chain is contained in a $\mathscr{C}$-chain, two general points of a fiber of τ can be connected by a $\mathscr{C}$-chain of length $\delta = \dim(X) - \dim(Y)$.

Lemma 5.10 *Let X^0 be a dense open subset of X, let Y^0 be a variety, and let $f : X^0 \to Y^0$ be a morphism with the property that two general points of a fiber of f can be connected by a $\mathscr{C}$-chain of length m'. There exists a diagram*

$$
\mathscr{C}' = \mathscr{C} \times_X X' \xrightarrow{\ F'\ } X' \xrightarrow{\ f'\ } Y
$$

where

- Y is a projective variety, Y^1 is a dense open subset of Y^0 and of Y, and $X^1 = f^{-1}(Y^1)$;

- X' is a projective variety and u is birational;

- f' is flat;

- any two *points in a fiber of f' can be connected by a $\mathscr{C}'$-chain of length m'.*

PROOF. Let Γ be the closure in $X \times Y^0$ of the graph of f. By generic flatness (5.4), there exists a dense open quasi-projective subset Y^1 of Y^0 over which the projection $\Gamma \to Y^0$ is flat (and proper). This yields a morphism $g : Y^1 \to \mathrm{Hilb}(X)$. Let $\overline{Y^1}$ be a compactification of Y^1 and let Y the closure of the graph of g in $\overline{Y^1} \times \mathrm{Hilb}(X)$. It is a projective variety that contains Y^1 as a dense open set, with a morphism $Y \to \mathrm{Hilb}(X)$ that extends g. Let $X' \to Y$ be the pull-back of the universal family; X' is a closed subscheme of $X \times Y$ that contains Γ as an open set. We have a diagram

$$
\begin{array}{ccccc}
\Gamma & \subset & X' & \subset & X \times Y \\
\downarrow & & \downarrow{\scriptstyle f'} & & \\
Y^1 & \subset & Y & \longrightarrow & \mathrm{Hilb}(X)
\end{array}
$$

where $f' : X' \to Y$ is projective and flat, and $f'^{-1}(Y^1) = \Gamma$.

Let Z' be an irreducible component of X'. Since f' is flat, we have $f'(Z') = Y$ (Lemma 5.3(a)), hence $f'^{-1}(Y^1) \cap Z'$ is dense in Z'. The latter being $\Gamma \cap Z'$, the irreducible open set Γ is dense in X', which is therefore irreducible.

By construction, u restricts to an isomorphism from $\Gamma \cap (X^1 \times Y^1)$ (which is the graph of f) onto X^1, hence is birational. We will identify these two open sets, making X^1 an open subset of X'.

Since f' is flat, so are the two projections $X' \times_Y X' \to X'$. Let Z be an irreducible component of $X' \times_Y X'$. By Lemma 5.3, it dominates each factor X', hence $Z \cap (X^1 \times X^1)$ is dense in Z. It follows that

$$
X^1 \times_{Y^1} X^1 = (X' \times_Y X') \cap (X^1 \times X^1)
$$

is dense in $X' \times_Y X'$. Let S be the subset of $X' \times X'$ that consists of pairs of points that can be connected by a $\mathscr{C}'$-chain of length m'. Since π and F are proper, S is a closed subscheme. Since a general pair of points of a fiber of f can be connected by a $\mathscr{C}$-chain of length m', it implies that *any* pair of points of the same fiber of f is contained in S. In other words, S contains $X^1 \times_{Y^1} X^1$, hence also its closure $X' \times_Y X'$. $\square$

By the third reduction, it is enough to find a quotient for $\mathscr{C}'$-equivalence on X'. The idea is to proceed by induction on the dimension of X, assuming that the quotient exists for the basic diagram

$$
\begin{array}{c}
\mathscr{C}' = \mathscr{C} \times_X X' \xrightarrow{\; f' \circ F' \;} Y \\
\downarrow{\scriptstyle \pi'} \\
T
\end{array}
$$

Lemma 5.11 *Keep the same hypotheses and notation as in Lemma 5.10, and assume that there is a proper quotient $\psi : Y^* \to Z^*$ for $\mathscr{C}'$-equivalence* *as in the theorem, where Y^* is open in Y. Set $X'^* = f'^{-1}(Y^*)$. If $m' < 2^{\dim(X)-\dim(Y)}$, the map $\psi \circ f' : X'^* \to Z^*$ is the required quotient for $\mathscr{C}'$-equivalence.*

PROOF. Since the notation is getting a bit out of control, I hope that the following diagram will help.

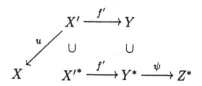

We may assume by the second reduction that π' has connected fibers. If the points x_1' and x_2' of X'^* are $\mathscr{C}'$-equivalent, so are $\tau'(x_1')$ and $\tau'(x_2')$, hence $\psi \circ f'(x_1') = \psi \circ f'(x_2')$ by property (a).

Conversely, if $\psi \circ f'(x_1') = \psi \circ f'(x_2')$, property (b) implies that $f'(x_1')$ and $f'(x_2')$ can be connected by a $\mathscr{C}'$-chain of length

$$m = 2^{\dim(Y)-\dim(Z^*)} - 1$$

because there exist points $t_1, \ldots, t_m$ such that $f'(x_1')$ is in $f'(F'(\mathscr{C}_{t_1}'))$, each $f'(F'(\mathscr{C}_{t_i}'))$ meets $f'(F'(\mathscr{C}_{t_{i+1}}'))$, and $f'(F'(\mathscr{C}_{t_m}'))$ contains $f'(x_2')$.

Therefore, $F'(\mathscr{C}_{t_1}')$ meets the fiber of x_1', the sets $F'(\mathscr{C}_{t_i}')$ and $F'(\mathscr{C}_{t_{i+1}}')$ have points that are in the same fiber of f', and $f'(F'(\mathscr{C}_{t_m}'))$ meets the fiber of x_2'.

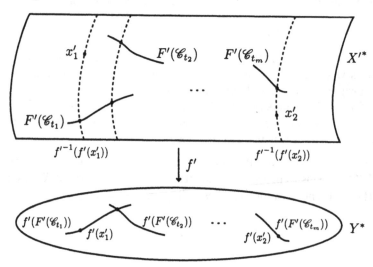

Connecting the points x_1' and x_2' of X'^* by a $\mathscr{C}'$-chain.

Since points in the same fiber of f' can be connected by a $\mathscr{C}$-chain of length m', it follows that x_1' and x_2' can be connected by a $\mathscr{C}'$-chain of length

$$m + (m+1)m'$$
$$\leq 2^{\dim(Y)-\dim(Z^*)} - 1 + 2^{\dim(Y)-\dim(Z^*)}(2^{\dim(X)-\dim(Y)} - 1)$$
$$= 2^{\dim(X)-\dim(Z^*)} - 1$$

We have proved that the proper map $\psi \circ f' : X'^* \to Z^*$ satisfies both properties (a) and (b) of the theorem. $\qquad\square$

We conclude by induction on the dimension of X and by Noetherian induction on $\mathscr{C}$. Let us go back to our quotient $\tau : X^0 \to Y^0$ for $\mathscr{C}^1$-equivalence. By Noetherian induction, there exists also a proper quotient $X_0 \to Y_0$ for $(\mathscr{C}-\mathscr{C}^1)$-equivalence. By Lemma 5.10, we may compactify these quotients, with

$$m' = \dim(X) - \dim(Y^0)$$

in the first case and
$$m' = 2^{\dim(X)-\dim(Y_0)} - 1$$

in the second case. If either $\dim(Y^0)$ of $\dim(Y_0)$ is less than $\dim(X)$, we can apply the induction hypothesis, and Lemma 5.11 gives the result.

Note that the proper morphism ρ is not necessarily a morphism of k-schemes, but can be made so upon replacing it with its Stein factorization.

If $\dim(X) = \dim(Y^0)$, any set-theoretic fiber of τ is a single point x by property (b) of Proposition 5.7. By property (a) of the same proposition, the closure of the $\mathscr{C}^1$-equivalence class of a point x of X^0 meets X^0 only at x, hence must consist only of x since it is connected. It follows that $\mathscr{C}^1$-equivalence is trivial on X^0: for any t in T, either $F(\mathscr{C}_t^1)$ lies entirely inside $X - X^0$, or it is a point.

If moreover $\dim(X) = \dim(Y_0)$, then similarly, for any t in T, either $F(\mathscr{C}_t-\mathscr{C}_t^1)$ lies entirely inside $X - X_0$, or it is a point. This implies that $\mathscr{C}$-equivalence is trivial on $X^0 \cap X_0$ and we can take the identity as the quotient map on this open subset of X. $\qquad\square$

5.5 Construction of the Rational Quotient

We have now all the material we need to construct the rational quotient. The only thing we have to deal with is the possibly varying degrees of the rational curves. This amounts to constructing a quotient for countably many "nondecreasing" basic diagrams (5.1). To make statements easier, we introduce a bit of terminology. Rationally chain-connected varieties are defined in Definition 4.21.

Recall that the ground field k is algebraically closed.

Definition 5.12 *Let X be a normal and proper variety.*

- *A fibration is a rational map $\rho : X \dashrightarrow Z$ that restricts on a dense open subset X^0 of X to a proper map $\rho^0 : X^0 \to Z$.*

- *A fibration $\rho : X \dashrightarrow Z$ that is defined and proper on a dense open set X^0 is rationally chain-connected if its fibers are rationally chain-connected and $\rho^0_* \mathcal{O}_{X^0} \simeq \mathcal{O}_Z$*

We may always assume that Z is normal, in which case, as explained in 1.13, the second condition is equivalent in characteristic zero to the connectedness of the fibers of π, hence follows from their rational connectedness. It is necessary only in the case of positive characteristic to avoid inseparability phenomena.

Recall that two points on a variety are $\mathcal{R}$-equivalent if they can be connected by a chain of rational curves.

Theorem 5.13 *Let X be a normal projective variety. There exist a normal variety $R(X)$ and a rationally chain-connected fibration $\rho : X \dashrightarrow R(X)$ such that for any other rationally chain-connected fibration $\psi : X \dashrightarrow Z$, there is a unique rational map $\pi : Z \dashrightarrow R(X)$ such that $\rho = \pi \circ \psi$.*

Very general fibers of ρ are $\mathcal{R}$-equivalence classes.

Recall that a property holds for a *very general* point of a variety if it holds for any point outside of a countable union of proper subvarieties. So the last statement of the definition is useful only for uncountable ground fields.

The variety $R(X)$ in the theorem is uniquely determined by X up to birational equivalence. We will (abusively) call it *the rational quotient of X*. A proper and normal variety is rationally chain-connected if and only if its rational quotient is a point.

Since X is normal, it follows from 5.5 that a very general $\mathcal{R}$-equivalence class is an *irreducible* subvariety of X. Also, the proof below shows that ρ is a quotient in the sense of Theorem 5.9 for some family of rational curves. It follows from that theorem that any two points of a very general fiber can be connected by a rational chain of length $2^{\dim(X)-\dim(R(X))} - 1$.

PROOF OF THE THEOREM. Recall that 1-dimensional connected closed subschemes of X with rational components are parametrized by a (locally Noetherian) scheme $\mathrm{Rat}(X)$, which is the union of a nondecreasing sequence $(\mathrm{Rat}_m(X))_{m \geq 0}$ of *projective* schemes (see 5.6). For each integer m, there is a diagram

$$\begin{array}{ccc} \mathscr{C}_m & \xrightarrow{\;F_m\;} & X \\ {\scriptstyle \pi_m}\downarrow & & \\ \mathrm{Rat}_m(X) & & \end{array}$$

where $\mathscr{C}_m$ is projective. Theorem 5.9 yields a dense open set X_m^0 of X and a proper map $\rho_m : X_m^0 \to Z_m$ whose fibers are $\mathscr{C}_m$-equivalence classes. Since $\mathscr{C}_m$-equivalence classes grow with m, the sequence $(\dim(Z_m))_{m \geq 0}$ is nonincreasing hence must stabilize for $m \geq m_0$. Using 5.5, this implies that for each $m \geq m_0$, there exists a dense open subset Z_m^0 of Z_m over which fibers of ρ_m are proper irreducible $\mathscr{C}_m$-equivalence classes of dimension $\dim(X) - \dim(Z_{m_0})$. Since $\mathscr{C}_m$-equivalence classes grow with m, it follows that $\mathscr{C}_m$- and $\mathscr{C}_{m_0}$-equivalence classes coincide on

$$X_m^1 = \rho_{m_0}^{-1}(Z_{m_0}^0) \cap \rho_m^{-1}(Z_m^0)$$

This implies that X_m^1 is a union of fibers of ρ_{m_0}, and that the fiber of ρ_{m_0} through any point x in $\bigcap_{m \geq m_0} X_m^1$ is the $\mathscr{R}$-equivalence class of x. This proves that the morphism ρ that appears in the Stein factorization

$$\rho_{m_0} : X_{m_0}^0 \xrightarrow{\rho} R(X) \to Z_{m_0}$$

is a rationally chain-connected fibration and that its very general fibers are $\mathscr{R}$-equivalence classes.

Let us prove that ρ is maximal. The idea is that since ρ contracts rational curves, it must in particular contract the fibers of any other rationally chain-connected fibration. Let $\psi : X \dashrightarrow Z$ be a fibration defined and proper over a dense open subset X^1 of X. We may assume that the base field is uncountable (let Y be the image of the map $(\rho, \psi) : X^0 \cap X^1 \to R(X) \times Z$; as explained in the proof of Lemma 1.15, the existence of the factorization π is equivalent to the birationality of the first projection $Y \to R(X)$, and this can be checked after a base extension).

Let x be a very general point of X^1. Since the fibers of ψ are rationally chain-connected, any point in $\psi^{-1}(\psi(x))$ is $\mathscr{R}$-equivalent to x, hence belongs to the same fiber of ρ. In other words, ρ is defined on and contracts $\psi^{-1}(\psi(x))$. Since ψ is proper over some neighborhood of x, there is an open neighborhood Z^0 of x in Z such that ρ is defined on $\psi^{-1}(Z^0)$. Since $\psi_* \mathscr{O}_{\psi^{-1}(Z^0)} \simeq \mathscr{O}_{Z^0}$, the factorization follows from Lemma 1.15. $\square$

The following result gives a characterization of the dimension of the rational quotient of a smooth projective variety (as defined in Theorem 5.13) in terms of the rational curves that it contains. It generalizes both Corollaries 4.11 and 4.17.

Corollary 5.14 *Assume the characteristic is zero. Let X be a smooth projective variety and let $X \dashrightarrow R(X)$ be its rational quotient. The dimension of $R(X)$ is the smallest integer r such that there exists a free rational curve $f : \mathbf{P}^1 \to X$ such that $f^* T_X$ has r trivial summands.*

PROOF. We may assume that the base field is uncountable. For a free rational curve $f : \mathbf{P}^1 \to X$, the property that $f^* T_X$ has (at least) r trivial summands is equivalent to $h^1(\mathbf{P}^1, f^* T_X(-2)) \leq r$, hence it still holds for a general (flat) deformation of f by semicontinuity ([H1], III, th. 12.8). So we may assume that f passes through a very general point of X. In that case, since very general fibers of the rational quotient $\rho : X \dashrightarrow R(X)$ are $\mathscr{R}$-equivalence classes, $f(\mathbf{P}^1)$ must be contracted to a point z. By generic smoothness, the fiber $F = \rho^{-1}(z)$ is smooth projective, and its normal bundle in X is trivial. The exact sequence

$$0 \to f^*T_F \to f^*T_X \to \mathcal{O}_{\mathbf{P}^1} \otimes T_{R(X),z} \to 0$$

implies $r \geq \dim(R(X))$.

Conversely, since F is smooth and rationally chain-connected, it is rationally connected by Corollary 4.28, hence contains a very free rational curve $f : \mathbf{P}^1 \to F$ by Corollary 4.11. By the same exact sequence, f^*T_X has at least $\dim(F)$ nontrivial summands, hence $r \leq \dim(R(X))$.   □

Example 5.15 We saw in Exercise 4.8.3 that if X is a smooth projective uniruled variety (still in characteristic zero), there is a *minimal* free rational curve $f : \mathbf{P}^1 \to X$. It satisfies

$$f^*T_X \simeq \mathcal{O}_{\mathbf{P}^1}(2) \oplus \mathcal{O}_{\mathbf{P}^1}(1)^{s-1} \oplus \mathcal{O}_{\mathbf{P}^1}^{n-s} \tag{5.2}$$

where s is the dimension of the (closed) subvariety of X swept out by free rational curves of minimal degree passing through a very general point x of X. This variety is contained in the fiber $\rho^{-1}(\rho(x))$ of the rational quotient $\rho : X \dashrightarrow R(X)$, hence $s \leq n - \dim(R(X))$, which is coherent with the conclusion $\dim(R(X)) \leq n - s$ of the corollary. The inequality is strict in general: in Section 5.11, we study the Fano variety

$$X = \mathbf{P}(\mathcal{O}_{\mathbf{P}^r}^s \oplus \mathcal{O}_{\mathbf{P}^r}(r))$$

It is rationally connected by Proposition 5.16, hence its rational quotient is a point. However, the lines in the fibers of $X \to \mathbf{P}^s$ are minimal and the pull-back of T_X is given by (5.2).

5.6   Rational Connectedness of Fano Varieties

As promised earlier, we use the rational quotient to prove that Fano varieties are rationally chain-connected, and rationally connected when the characteristic is zero. This is a result due to Campana ([C3]), and Kollár, Miyaoka and Mori ([KMM3]). We even get explicit bounds on the degrees of the rational curves involved.

Proposition 5.16 *A Fano variety is rationally chain-connected. More precisely, any two points of a Fano variety X of dimension n can be connected by a chain of rational curves of total $(-K_X)$-degree at most $(2^n - 1)(n+1)$.*

Moreover, if the characteristic is zero, any two points of X can be connected by a single very free rational curve of $(-K_X)$-degree at most

$$3(2^n - 1)(n + 1)^{(n+1)(2^n - 1)}$$

When X has Picard number 1, we get the bound $2(n + 1)^{n^2+n+1}$ in the second point, because we may take $d = n(n + 1)$ in the proof by Proposition 5.8. But the much better bound of Proposition 5.8, which says that a *general* pair of points can be connected by a very free rational curve of $(-K_X)$-degree at most $n(n + 1)$ (see, however, footnote 6, p. 121), will be sufficient for the applications we have in mind.

We will construct in Section 5.11, for each integer n, a Fano variety X of dimension n for which the minimal $(-K_X)$-degree of any irreducible curve joining a point of X to a general point is $O(n^{\frac{\log n}{\log 4}})$.

As to the second point, what is known in positive characteristic? Not much: the results proved here make essential use of the existence of a free rational curve on a Fano variety, which is not true in general ([K1], §V.5).

PROOF OF THE PROPOSITION. Since rational chain-connectedness does not depend on the field over which X is defined (Remark 4.22(3)), we may assume that it is uncountable. Let $\rho : X \dashrightarrow R(X)$ be the rational quotient. If $R(X)$ is not a point, there exists by Theorem 3.22 a rational curve not contracted by ρ through a point of a very general fiber of ρ. This is absurd since the latter is an $\mathscr{R}$-equivalence class. It follows that $R(X)$ is a point and X is rationally chain-connected.

Any two points of X can therefore be connected by a rational chain, whose links can be broken up by Proposition 3.2 until they have $(-K_X)$-degree at most $n+1$. Let $\mathrm{Rat}_{n+1}(X)$ be the closure in $\mathrm{Hilb}(X)$ of the scheme that parametrizes curves in X with rational components, of total $(-K_X)$-degree at most $n + 1$ (see 5.6), and let $\mathscr{C} \to \mathrm{Rat}_{n+1}(X)$ be the universal family. There is only one equivalence class for $\mathscr{C}$-equivalence, hence any two points of X can be connected by a $\mathscr{C}$-chain of length at most $2^n - 1$ by Theorem 5.9. This proves the first point.

Assume now that the characteristic is zero. By Proposition 5.16, the Theorem 4.27 applies with $H = -K_X$ and $d = (n + 1)(2^n - 1)$. Since $K_X \cdot C_{i+1} < 0$, we may take $m = n + 1$ in (4.6) and the estimate (4.7) yields the better bound $2d(n + 1)^d/n$ on the degree.   □

5.7   Bounds on the Degree of Fano Varieties

We now know from Proposition 5.16 that any two points on a Fano variety X can be joined by a rational curve with some universal bound on the degree of that curve. An elementary argument shows that under these circumstances, the self-intersection number $(-K_X)^n$ is also bounded by some constant that depends only on n (the better result of Proposition 5.8 gives a very good bound in case the Picard number is 1).

Proposition 5.17 *Let X be a projective variety of dimension n, let H be a nef divisor on X, and let d be a nonnegative real number. Let x be a smooth point of X such that a general point of X can be connected to x by an irreducible curve of H-degree at most d. We have $H^n \le d^n$.*

Proof. We may assume that H is big since otherwise there is nothing to prove. By Proposition 1.31(b), $h^0(X, mH)$ is equivalent to

$$\frac{H^n}{n!} m^n = \frac{(m \sqrt[n]{H^n})^n}{n!}$$

for $m \gg 0$. Let $\mathfrak{m}_{X,x}$ be the maximal ideal of $\mathcal{O}_{X,x}$, and let r be a positive integer. Because of the exact sequence

$$0 \to H^0(X, \mathfrak{m}^r_{X,x}(mH)) \to H^0(X, mH) \to H^0(X, \mathcal{O}_{X,x}/\mathfrak{m}^r_{X,x}) \simeq k^{\binom{n+r-1}{n}}$$

having a point of multiplicity at least r imposes at most $\binom{n+r-1}{n} \sim \frac{r^n}{n!}$ conditions on the elements of the linear system $|mH|$. It follows that given $\varepsilon > 0$, there exist a large positive integer m and a divisor D in $|mH|$ such that

$$\mathrm{mult}_x D \ge m \sqrt[n]{H^n} - m\varepsilon$$

Let C be an irreducible curve of H-degree at most d connecting x to a point outside of D. Since C is not contained in the support of D, we have (this is the point where having an irreducible curve, instead of merely a chain, connecting x and this other point, is essential)

$$md \ge C \cdot (mH) = C \cdot D \ge \mathrm{mult}_x D \ge m \sqrt[n]{H^n} - m\varepsilon$$

and the proposition follows by letting ε go to 0.   □

Theorem 5.18 *Assume the characteristic is zero. For any Fano variety X of dimension n,*

$$(-K_X)^n \le (3(2^n - 1)(n + 1)^{(n+1)(2^n-1)})^n$$

If X has Picard number 1,

$$(-K_X)^n \le (n(n + 1))^n$$

This type of bound was first established by Batyrev in [Ba] for *toric* Fano varieties.

Proof of the Theorem. The general case is a consequence of Propositions 5.17 and 5.16, and the case when X has Picard number 1 of Propositions 5.17 and 5.8.   □

Since

$$(-K_{\mathbf{P}^n})^n = (n+1)^n$$

the second bound of the theorem is rather sharp. The first (general) bound is obviously much too large. Examples in Section 5.11 suggest that for a Fano variety X of dimension n and Picard number k, the degree $(-K_X)^n$ should be a $O(n^{kn})$. Bounding the number k in terms of n is another story, though. Examples in Section 5.11 show that k may grow as $\log n$.

When X has Picard number 1 and the characteristic is zero, Ran and Clemens obtain in [RC] a bound on the degree $(-K_X)^n$ which involves the *Fano index* ι_X of X, the greatest integer by which the class of K_X is divisible.[8] Their bound is

$$(-K_X)^n \leq (\max(\iota_X n, n+1))^n$$

Since the index is at most $n+1$,[9] it is stronger than the second bound of the theorem.

5.8   The Differential–Geometric Point of View

By Yau's proof of Calabi's conjecture ([Y1], [Y2]),[10] complex Fano varieties are compact Kähler varieties with positive definite Ricci curvature ([Bou1], [Be], 11.16(ii)). This leads to another proof of their simple-connectedness ([Be], th. 11.26). In particular, compact complex varieties with a positive Kähler–Einstein metric[11] are Fano varieties. However, not all complex Fano varieties are of this type, because this forces the automorphism group to be reductive[12] ([Be], cor. 11.54). This excludes for example the projective plane blown up at one or two points (see Exercise 5.12.1).[13] For these varieties however, classical methods of differential geometry yield a good bound on the degree.

---

[8]Beware! This Fano index is not to be confused with the index j_X of a variety X as defined in 7.14.

[9]This holds for any Fano variety by Theorem 3.4. There is also a neat proof in characteristic 0: write $K_X = \iota_X H$, with H ample. By Kodaira's vanishing theorem, the polynomial $P(m) = \chi(X, \mathcal{O}_X(K_X + mH))$ is equal to $h^0(X, \mathcal{O}_X(K_X + mH))$ for all positive integers m. Since it has degree n, it must be nonzero for some $m_0 \in \{1, \ldots, n+1\}$, and $-\iota_X \leq m_0$. It has been shown that the index is $n+1$ if and only if X is isomorphic to $\mathbf{P}^n$. It is n if and only if X is a smooth quadric.

[10]If (X, ω) is a compact Kähler manifold, its Ricci form is closed and its class is the first Chern class of X. Yau's theorem says that given a closed form ρ representing $c_1(X)$, there exists a Kähler metric on X whose Ricci form is ρ.

[11]This means that the Ricci curvature is a positive (constant) multiple of the metric.

[12]An algebraic group is reductive if it has no nontrivial connected unipotent abelian normal subgroups.

[13]For more results on the existence of a Kähler–Einstein metric on a Fano manifold, see [Bou2].

Let X be a compact complex variety of dimension n with a Kähler–Einstein metric g normalized so that its Ricci curvature is g. By [Be], 11.5, we have

$$(-K_X)^n = \text{vol}(X)\, \frac{n!}{(2\pi)^n}$$

From [BC], cor. 4, p. 257, it follows that the volume of X is less than the volume of a sphere of the same dimension and radius $\sqrt{2n-1}$, so that

$$(-K_X)^n \;\leq\; \frac{n!}{(2\pi)^n}\,(\sqrt{2n-1})^{2n}\,\frac{2^{n+1}\pi^n}{1\cdot 3\cdots(2n-1)}$$

$$= (2n-1)^n\,\frac{2^{n+1}(n!)^2}{(2n)!}$$

Let $d_{KE}(n)$ be this bound. It is equivalent to $2^{n+1}\sqrt{\frac{\pi n}{e}}\,n^n$. The existence of a Kähler–Einstein metric on X implies the stability[14] of the tangent bundle of X ([Lu]). If the Picard number of X is 1 and its tangent bundle is semistable, the slightly better bound

$$(-K_X)^n \leq (2n)^n$$

is proved in [RC].

5.9 Fano Varieties Form a Limited Family

It is out of the question to prove anything here. We will just give a few references.

There are only finitely many deformation types of Fano varieties X with fixed Hilbert polynomial $P(m) = \chi(X, \mathcal{O}_X(mK_X))$ (see [Mat]). Using the Hodge index theorem and the Kodaira vanishing theorem, one proves that in characteristic zero, for any ample divisor H on X, the coefficients of the Hilbert polynomial $\chi(X, \mathcal{O}_X(mH))$ are bounded by a function which depends only on the dimension n of X, on H^n and on $K_X \cdot H^{n-1}$ (see [K1], ex. VI.2.15.8). When $H = -K_X$, these numbers are bounded by a function that depends only on n by Theorem 5.18, hence there are finitely many deformation types of Fano varieties in any given dimension. One can even get an effective bound.

Theorem 5.19 *Let* **k** *be an algebraically closed field of characteristic zero. There are at most $(n+2)^{(n+2)^{n^{2^{3n}}}}$ deformation types of Fano varieties of dimension n defined over* **k***.*

---

[14]This means that for any proper subbundle $\mathcal{E}$ of T_X, we have

$$\frac{c_1(\mathcal{E}) \cdot (-K_X)^{n-1}}{\text{rank}(\mathcal{E})} < \frac{(-K_X)^n}{n}$$

This result was first proved in [N] (by a different method) and [KMM1], for Fano varieties with Picard number 1.

PROOF OF THE THEOREM. Let X be an n-dimensional Fano variety. It follows from [De1], [K3], [AS], and [T] that $n(n + 1)(n + 3)(-K_X)$ is very ample. By projection, X can therefore be realized as a smooth subvariety of $\mathbf{P}^{2n+1}$ of degree at most

$$\left(n(n + 1)(n + 3)(-K_X)\right)^n$$
$$\leq \left(3n(n + 1)(n + 3)(2^n - 1)(n + 1)^{(n+1)(2^n-1)}\right)^n = f(n)$$

by Theorem 5.18. The number of irreducible components of the Hilbert scheme of $\mathbf{P}^N$ parametrizing *smooth* subvarieties of dimension n and degree d is bounded by ([Ca] or [K1] (3.28.9), p. 61)

$$C(n, d, N) = \binom{d(N + 1)}{N}^{(N+1)d\binom{d+n}{n}}$$

The number of deformation types of Fano varieties of dimension n is therefore bounded by $C(n, f(n), 2n + 1)$, which is smaller than the bound given in the theorem. $\qquad\square$

In the differential-geometric context, LeBrun and Salamon give in [LS] a nice application of this finiteness result: they prove that there are, in each (real) dimension $4n$, and up to isometry and homothety, only finitely many compact "quaternionic-Kähler" (this means that their holonomy is contained in the subgroup $\mathrm{Sp}(n)\,\mathrm{Sp}(1)$ of $\mathrm{SO}(4n)$) Riemannian manifolds with positive scalar curvature. Their starting point is that the twistor space attached to such a variety is a complex Fano manifold with an extra structure called a contact structure (and with a Kähler–Einstein metric). The only known examples of quaternionic-Kähler varieties are homogeneous symmetric spaces called Wolf spaces, and it is conjectured that there are no others. This has been proved in real dimensions 4 and 8 as a consequence of results of [Dr].

Remarks 5.20 (1) A Fano curve is isomorphic to $\mathbf{P}^1$.

(2) A Fano surface is called a Del Pezzo surface. There are 10 families: $\mathbf{P}^2$ blown up in at most 8 points, and $\mathbf{P}^1 \times \mathbf{P}^1$ (this holds in any characteristic; see [K1], ex. III.3.9).

(3) Fano threefolds have been classified in characteristic zero: there are 17 families with Picard number 1 ([I1], [I2], [Sh2]), and 87 other families with Picard number > 1 ([MM]). In positive characteristic, see [SB], where Fano threefolds with Picard number 1 are classified, and [Me] for additional partial results.

(4) In view of Mori's minimal model program (see 7.7), it would be very interesting to have similar results for mildly singular Fano varieties,

more specifically, for what we will call **Q**-Fano varieties (see 7.42 for the definition).[15] Examples in the next section show that some restrictions on the singularities are necessary.

5.10   Singular Fano Varieties

For Mori's classification program, it is important (see for instance 7.45) to consider *singular* Fano varieties. By this, we mean a projective normal variety X such that some multiple of the Weil anticanonical divisor $-K_X$ (see footnote 3, p. 157) is Cartier and ample. As we remarked after Theorem 3.6, such a variety is uniruled by rational curves of $(-K_X)$-degree at most $2\dim(X)$.

Here is an example: let C be a smooth curve of genus g and degree d in $\mathbf{P}^n$ and let X be the cone over C in $\mathbf{P}^{n+1}$. There is a diagram

$$\tilde{X} = \mathbf{P}(\mathscr{O}_C \oplus \mathscr{O}_C(1)) \xrightarrow{\;f\;} X$$
$$\downarrow \pi$$
$$C$$

where the morphism f contracts the section C_0 of π associated with the trivial quotient of $\mathscr{O}_C \oplus \mathscr{O}_C(1)$ (one has $C_0^2 = -d$). The group $N_1(\tilde{X})$ has rank 2 and is generated by the class ℓ of a fiber of π and the class of the curve $C_1 = f^{-1}(C)$. On X, the hyperplane section C is linearly equivalent to the sum of d lines, hence $N_1(X)_\mathbf{R}$ has rank 1 and is generated by the (ample) class $f_*\ell$ of a line. One has ([H1], V, cor. 2.11)

$$K_{\tilde{X}} \sim -2C_0 + (2g - 2 - d)\ell$$

hence

$$K_X \sim (2g - 2 - d)f_*\ell$$

The hyperplane section of X has class $df_*\ell$. When $d > 2g - 2$, it follows that $-dK_X$ is an ample Cartier divisor, hence X is a *singular* Fano variety.

When g is positive, the rational curves on X produced by Theorem 3.6 must be the lines on $\tilde{X}$. They all meet its singular point, hence are not free in any sense (such as in Proposition 4.8). For further reference, note that

---

[15]In this direction, complex **Q**-Fano threefolds with canonical singularities (see Definition 7.13) and Picard number 1 are known to form a limited family ([Ka4] for the case of terminal singularities; the canonical case is unpublished work by Kollár, Miyaoka and Mori). Recent work of Ran and Clemens ([RC]) shows that **Q**-Fano n-folds with canonical singularities and Picard number 1 form a limited family if one moreover bounds the smallest number m such that $-mK_X$ is very ample (in the smooth case, one can take $m = n(n+1)(n+3)$ as in the proof of Theorem 5.19).

the number α such that $K_{\tilde{X}} = f^* K_X + \alpha C_0$ can be computed by taking intersections with C_0. We get

$$K_{\tilde{X}} = f^* K_X - \frac{2g - 2 + d}{d} C_0$$

Also, this construction yields infinitely many families of singular Fano surfaces, which shows that Theorem 5.19 is no longer valid in the singular case. It is expected that by limiting the singularities, there might still be only finitely many families.[16]

5.11 Examples of Fano Varieties with High Degree

Batyrev remarked that for the n-dimensional Fano variety

$$X = \mathbf{P}\big(\mathscr{O}_{\mathbf{P}^{n-1}} \oplus \mathscr{O}_{\mathbf{P}^{n-1}}(n - 1) \big)$$

where as usual, we follow Grothendieck's notation (for a vector bundle $\mathscr{E}$, the projectivization $\mathbf{P}(\mathscr{E})$ is the space of *hyperplanes* in the fibers of $\mathscr{E}$), we have

$$(-K_X)^n = \frac{(2n - 1)^n - 1}{n - 1} \sim \frac{2^n e^{-3/2}}{n} (n + 1)^n$$

We can do better by changing the numbers in the construction.

Proposition 5.21 *For each $n \geq 2$, there is a Fano variety X of dimension n, index 1 and Picard number 2 such that*

$$(-K_X)^n \geq \left(\frac{3n^2}{10 \log n} \right)^n$$

PROOF. Let

$$X = \mathbf{P}(\mathscr{O}_{\mathbf{P}^s}^r \oplus \mathscr{O}_{\mathbf{P}^s}(a))$$

where r, s and a are nonnegative integers. We have

$$-K_X \sim (r + 1)L + (s + 1 - a)H$$

where L is a divisor associated with the line bundle $\mathscr{O}_X(1)$ and H is the pull-back on X of a hyperplane in $\mathbf{P}^s$. It follows that X is a Fano variety

---

[16]This should be the case for example for Fano varieties with *canonical* singularities, as defined in Definition 7.13, and has been proved in dimension 3 in [Ka4] (the point is again to bound $(-K_X)^n$ and an integer j such that jK_X is Cartier). More generally, it is conjectured that given $\varepsilon > 0$, Fano varieties X such that, for any desingularization $f : \tilde{X} \to X$, one has $K_{\tilde{X}} = f^* K_X + \sum a_i E_i$ with $a_i > -1 + \varepsilon$, where the E_i are the exceptional divisors of f, still form a limited family. Our surfaces fall in this class for $g = 0$ and $d < 2/\varepsilon$.

when $a \leq s$ (this can either be seen directly by applying Kleiman's criterion 1.27(a), or by noting that $\mathcal{O}_X(-K_X)$ is the tautological line bundle $\mathcal{O}(1)$ on X associated with the description of X as $\mathbf{P}(\mathcal{O}_{\mathbf{P}^s}(s+1-a)^r \oplus \mathcal{O}_{\mathbf{P}^s}(s+1))$; it is therefore ample) .

In the intersection ring of X, we have the relation

$$L^{r+1} = aH \cdot L^r$$

which comes from the fact that $\mathcal{O}_X(1)$ is a quotient of the pull-back of the vector bundle $\mathcal{O}_{\mathbf{P}^s}^r \oplus \mathcal{O}_{\mathbf{P}^s}(a)$; hence the $(r+1)$-st Chern class of the kernel vanishes, and

$$L^r \cdot H^s = 1$$

Setting $n = \dim(X) = r + s$, we get for $a = s$

$$
\begin{aligned}
(-K_X)^n &= \sum_{i=0}^{n} \binom{n}{i} (r+1)^i L^i \cdot H^{n-i} \\
&= \sum_{i=r}^{n} \binom{n}{i} (r+1)^i (aH)^{i-r} \cdot L^r \cdot H^{n-i} \\
&= \sum_{i=r}^{n} \binom{n}{i} (r+1)^i a^{i-r} \\
&\geq (r+1)^n a^{n-r}
\end{aligned}
$$

Take $a = s = n - r$. The function $r \mapsto r^n(n-r)^{n-r}$ reaches its maximum near $\frac{n}{\log n}$. Taking $r = \left[\frac{n}{\log n}\right]$, we get

$$
\begin{aligned}
(-K_X)^n &\geq \left(\frac{n}{\log n}\right)^n \left(n - \frac{n}{\log n}\right)^{n - \frac{n}{\log n}} \\
&\geq n^{2n - \frac{n}{\log n}} \frac{1}{(\log n)^n} \left(1 - \frac{1}{\log n}\right)^n \\
&= n^{2n} e^{-n} \frac{1}{(\log n)^n} \left(1 - \frac{1}{\log n}\right)^n \\
&\geq \left(\frac{3n^2}{10 \log n}\right)^n
\end{aligned}
$$

for $\log n \geq \frac{10}{10-3e}$, i.e., $n \geq 226$. This lower bound for $(-K_X)^n$ actually holds for all $n \geq 3$ by direct calculation. $\square$

Note that the right-hand side of the inequality in the proposition is larger than the bound $d_{KE}(n)$ for n large (with the best choice of r in the construction, this is the case for $n \geq 6$). This is not surprising since the automorphism group of X is not reductive (there is a surjective morphism $\mathrm{Aut}(X) \to \mathrm{PGL}(s+1, \mathbf{C}) \times \mathrm{GL}(r, \mathbf{C})$ whose kernel is isomorphic to $H^0(\mathbf{P}^s, \mathcal{O}_{\mathbf{P}^s}(s-r))^r$ and consists of unipotent elements).

If we analyze the construction, we see that we started from a Fano variety with both high index and high degree. The variety X constructed in the proposition has index 1, hence cannot be used to repeat the process. However, if one takes instead $a = s - r = n - 2r$ and the same r, the index of X becomes $r + 1$, and, although the degree becomes slightly smaller, it still satisfies

$$(-K_X)^n \geq \left(\frac{n}{\log n}\right)^n \left(n - 2\frac{n}{\log n}\right)^{n - \frac{n}{\log n}}$$

$$\geq \left(\frac{n^2}{7 \log n}\right)^n$$

for $\log n \geq \frac{14}{7-e}$, but this lower bound is actually valid for all $n \geq 4$.

Proposition 5.22 *For each positive integers $k \geq 2$ and $n \geq 4$ such that $\frac{n}{\log n} \geq 2^{k-2}$, there exists a Fano variety X of dimension n and Picard number k such that*

$$(-K_X)^n \geq \left(\frac{n^k}{2^{k^2-1}(\log n)^{k-1}}\right)^n$$

If one chooses k to be the largest integer such that 2^{2^k} is smaller than $\frac{n}{\log n}$, so that $k \sim \frac{\log n}{2 \log 2}$, we get for n large

$$(-K_X)^n \geq \left(\frac{n}{2^{k+1} \log n}\right)^{n(k-1)} \sim n^{\frac{n \log n}{4 \log 2}}$$

In particular, *there is no upper bound on $\sqrt[n]{(-K_X)^n}$ that is polynomial in n.*

PROOF OF THE PROPOSITION. We proceed by induction on k, *assuming in addition that the index of X is $\left[\frac{n}{2^{k-2} \log n}\right] + 1$. We just did it for $k = 2$.* Assume the construction is done for some $k \geq 2$. Let n be an integer as in the proposition, and set

$$r = \left[\frac{n}{2^{k-1} \log n}\right] \quad \text{and} \quad s = n - r$$

Since $\frac{n}{\log n} \geq 2^{k-1}$, the integer r is positive. Also, $r \leq \frac{n}{4}$ because $n \geq 4$ and $k \geq 2$. It implies

$$\frac{s}{\log s} \geq \frac{3n}{4 \log n} > 2^{k-2}$$

hence there exists by induction a Fano variety Y of dimension s, Picard number k, and index $\iota_Y = \left[\frac{s}{2^{k-2} \log s}\right] + 1$ such that

$$(-K_Y)^s \geq \left(\frac{s^k}{c_k (\log s)^{k-1}}\right)^s$$

for some positive constant c_k. Write $-K_Y = \iota_Y H$, with H ample on Y, and let

$$X = \mathbf{P}(\mathcal{O}_Y^r \oplus \mathcal{O}_Y((\iota_Y - r - 1)H))$$

with projection $\pi : X \to Y$, so that $-K_X = (r+1)(L + \pi^* H)$. Since

$$\iota_Y \geq \frac{s}{2^{k-2} \log s} \geq \frac{3n}{2^k \log n} > \frac{n}{2^{k-1} \log n} \geq r$$

X is a Fano variety of dimension $n = r + s$ with $\rho_X = k+1$ and $\iota_X = r+1$. We get again

$$
\begin{aligned}
(-K_X)^n &= (r+1)^n \sum_{i=r}^{n} \binom{n}{i} (\iota_Y - r - 1)^{i-r} H^s \\
&\geq (r+1)^n \left(\binom{n}{r} + (\iota_Y - r - 1)^s \right) H^s \\
&\geq \left(\frac{n}{2^{k-1} \log n} \right)^n (1 + (\iota_Y - r - 1)^s) \left(\frac{c_k s^k}{(\log s)^{k-1}} \right)^s \frac{1}{\iota_Y^s}
\end{aligned}
$$

Note that

$$\iota_Y \geq \frac{s}{2^{k-2} \log n} \geq \frac{3n}{2^k \log n}$$

If $\frac{n}{\log n} \geq 7 \cdot 2^{k-2}$, we obtain

$$\frac{r+1}{\iota_Y} \leq \frac{2}{3} + \frac{2^k \log n}{3n} \leq \frac{6}{7}$$

If $\frac{n}{\log n} < 7 \cdot 2^{k+1}$, we get

$$\frac{1}{\iota_Y} \geq \frac{1}{\frac{s}{2^{k-2} \log s} + 1} \geq \frac{1}{\frac{n}{2^{k-2} \log n} + 1} \geq \frac{1}{8}$$

In all cases,

$$(1 + (\iota_Y - r - 1)^s) \frac{1}{\iota_Y^s} \geq \frac{1}{8^n}$$

It follows that

$$
\begin{aligned}
((-K_X)^n)^{1/n} &\geq \frac{n}{2^{k-1} \log n} \left(\frac{c_k (3n)^k}{4^k (\log n)^{k-1}} \right)^{1-\frac{r}{n}} \frac{1}{8} \\
&\geq n^{1+k-k\frac{1}{2^{k-1} \log n}} \frac{1}{2^{k+2} \log n} \frac{c_k 3^k}{4^k (\log n)^{k-1}} \\
&\geq \frac{c_{k+1} n^{k+1}}{(\log n)^k}
\end{aligned}
$$

where $c_{k+1} = \dfrac{c_k 3^k}{4e8^k}$. This yields (since we may begin with $c_2 = 1/7$)

$$c_k = \frac{1}{7(4e)^{k-2}} \left(\frac{3}{8}\right)^{(k-1)(k-2)/2} \geq \frac{1}{2^{k^2-1}}$$

and proves the proposition.   □

Using Proposition 5.17, this implies that for any point x on X, the minimal $(-K_X)$-degree $\delta_X(x)$ of an *irreducible* curve connecting x to a general point of X is at least $\dfrac{n^k}{2^{k^2-1}(\log n)^{k-1}}$ for the varieties constructed above. Using the same construction with slightly different numbers, one can even get

$$\delta_X(x) \geq \frac{n^k}{2^{k(k+7)/2-6}} \tag{5.3}$$

This is done as follows: to the trivial rank r quotient of the vector bundle defining X is associated a divisor E in X isomorphic to $\mathbf{P}^{r-1} \times Y$ and numerically equivalent to $L - (\iota_Y - r - 1)\pi^* H$. For any irreducible curve C not contained in E, we have $E \cdot C \geq 0$, hence $-K_X \cdot C \geq (r+1)(\iota_Y - r)H \cdot \pi_* C$ and

$$\delta_X(x) \geq (r+1)\left(1 - \frac{r}{\iota_Y}\right)\delta_Y(\pi(x)) \tag{5.4}$$

(one proves that x can actually be connected to a general point of X by an irreducible *rational* curve C such that $E \cdot C = 0$, hence (5.4) is actually an equality). Taking $r = \left[\frac{n}{2^{k+2}}\right]$ in the proof of the proposition yields the (5.3).

5.12   Exercise

1. Show that the automorphism group of the plane $\mathbf{P}^2$ blown up at one point is isomorphic to the group of matrices of the form

$$\begin{pmatrix} 1 & \star & \star \\ 0 & \star & \star \\ 0 & \star & \star \end{pmatrix}$$

and that the set of matrices of the form

$$\begin{pmatrix} 1 & \star & \star \\ 0 & 1 & 0 \\ 0 & 0 & 1 \end{pmatrix}$$

is a unipotent abelian normal subgroup. This group is therefore not reductive. Prove an analogous result for the plane blown up at two points.

6

The Cone of Curves in the Smooth Case

Let X be a smooth projective variety. We defined in Chapter 1 the cone of curves $\mathrm{NE}(X)$ of X as the convex cone in $N_1(X)_{\mathbf{R}}$ generated by classes of effective curves. We prove here Mori's theorem on the structure of the closure $\overline{\mathrm{NE}}(X)$ of this cone, more exactly of the part where K_X is negative. We show that it is generated by countably many *extremal rays*, and that these rays can only accumulate on the hyperplane $K_X = 0$. We will give in the next chapter (§7.9) a totally different proof of the cone theorem which works for mildly singular varieties, but only in characteristic zero, and relies heavily on cohomology calculations. The methods of this second proof will also give a very important additional piece of information: the existence of the contraction (see 1.16) of extremal rays on which K_X is negative, which is at present unattainable by Mori's geometric approach.

Mori's method is still worthwhile studying, if only because it works in any characteristic, but also because it is such a beautiful application of his bend-and-break results (more precisely of Theorem 3.6).

We state the theorem in Section 6.1, but before giving its proof, we explain its geometric significance in Section 6.2 on some examples (mainly surfaces). These examples will also show that the structure of the cone $\overline{\mathrm{NE}}(X)$ can be quite complicated.

In Section 6.3, we prove elementary properties of closed convex cones in $\mathbf{R}^n$, their extremal subcones, and their supporting hyperplanes.

The cone theorem is proved in Section 6.4. In Section 6.5, we study geometric properties of the contraction $X \to Y$ of an extremal ray on which K_X is negative (without proving its existence). They are classified in three categories: fiber contractions (the general fiber is positive-dimensional), divisorial contractions (the exceptional locus is a divisor), and small contractions (the exceptional locus has codimension at least 2). Small contractions are the most difficult to handle: in this case, Y is too singular and the minimal model program can only continue if one can construct a *flip* $X^+ \to Y$ of the contraction $X \to Y$ (see 6.13). The existence of a flip is still unknown in general (see Section 7.12). Examples of contractions and flips are given in Section 6.6.

Everything takes place over an algebraically closed field $\mathbf{k}$.

6.1   Statement of the Cone Theorem

As promised, we state the cone theorem for smooth projective varieties.

If X is a proper scheme, D a Cartier divisor on X, and S a subset of $N_1(X)_{\mathbf{R}}$, we set

$$S_{D \geq 0} = \{z \in S \mid D \cdot z \geq 0\}$$

and similarly for $S_{D \leq 0}$, $S_{D > 0}$ and $S_{D < 0}$.

Theorem 6.1 *Let X be a smooth projective variety. There exists a countable family $(\Gamma_i)_{i \in I}$ of rational curves on X such that*

$$0 < -K_X \cdot \Gamma_i \leq \dim(X) + 1$$

and

$$\overline{\mathrm{NE}}(X) = \overline{\mathrm{NE}}(X)_{K_X \geq 0} + \sum_{i \in I} \mathbf{R}^+[\Gamma_i]$$

where the $\mathbf{R}^+[\Gamma_i]$ are all the (distinct) extremal rays of $\overline{\mathrm{NE}}(X)$ that meet $N_1(X)_{K_X < 0}$. These rays are locally discrete in that half-space.

An extremal ray that meets $N_1(X)_{K_X < 0}$ is called K_X-*negative*. Proposition 1.45 already showed the importance of rational curves Γ on X such that $K_X \cdot \Gamma < 0$ for studying birational morphisms defined on X which are not isomorphisms.

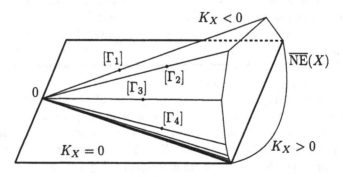

The closed cone of curves.

6.2 Examples: The Cone Theorem in Action

Before proving the cone theorem, we will illustrate it by a series of examples having to do mainly with surfaces. We first prove a lemma that will help us locate extremal curves on the closed cone of curves of a projective surface.

Lemma 6.2 *Let X be a smooth projective surface.*

(a) *The class of an irreducible curve C on X satisfying $C^2 \leq 0$ is in $\partial \overline{NE}(X)$.*

(b) *The class of an irreducible curve C on X satisfying $C^2 < 0$ spans an extremal ray of $\overline{NE}(X)$.*

(c) *If the class of an irreducible curve C on X satisfying $C^2 = 0$ and $K_X \cdot C < 0$ spans an extremal ray of $\overline{NE}(X)$, the surface X is ruled over a smooth curve, C is a fiber and X has Picard number 2.*

(d) *If r spans an extremal ray of $\overline{NE}(X)$, either $r^2 \leq 0$ or X has Picard number 1.*

(e) *If r spans an extremal ray of $\overline{NE}(X)$ and $r^2 < 0$, the extremal ray is spanned by the class of an irreducible curve.*

If X has Picard number at least 3, the cone theorem and the lemma imply that any K_X-negative extremal ray is spanned by an exceptional curve.

PROOF OF THE LEMMA. Assume $C^2 = 0$. Then $[C]$ has nonnegative intersection with the class of any effective divisor, hence with any element of $\overline{NE}(X)$. Let H be an ample divisor on X. If $[C]$ is in the interior of $\overline{NE}(X)$, so is $[C] + t[H]$ for all t small enough. This implies

$$0 \leq C \cdot (C + tH) = tC \cdot H$$

for all t small enough, which is absurd since $C \cdot H > 0$.

Assume now $C^2 < 0$ and $[C] = z_1 + z_2$, where z_i is the limit of a sequence of classes of effective divisors $D_{i,m}$. Write

$$D_{i,m} = a_{i,m} C + D'_{i,m}$$

with $a_{i,m} \geq 0$ and $D'_{i,m}$ effective with $C \cdot D'_{i,m} \geq 0$. Taking intersections with H, we see that the upper limit of the sequence $(a_{i,m})_m$ is at most 1, so we may assume that it has a limit a_i. In that case, $([D'_{i,m}])_m$ also has a limit $z'_i = z_i - a_i[C]$ in $\overline{NE}(X)$ which satisfies $C \cdot z'_i \geq 0$. We have then $[C] = (a_1 + a_2)[C] + z'_1 + z'_2$, and by taking intersections with C, we get $a_1 + a_2 \geq 1$. But

$$0 = (a_1 + a_2 - 1)[C] + z'_1 + z'_2$$

and since X is projective, this implies $z'_1 = z'_2 = 0$ and proves (b) and (a).

Let us prove (c). By the adjunction formula ([H1], V, ex. 1.3), we have $K_X \cdot C = -2$ and C is smooth rational.

For any divisor D on X such that $D \cdot H > 0$ and $m > \frac{K_X \cdot H}{D \cdot H}$, the divisor $K_X - mD$ has negative intersection with H, hence cannot be equivalent to an effective divisor. It follows that $H^0(X, \mathscr{O}_X(K_X - mD))$ vanishes, hence

$$H^2(X, \mathscr{O}_X(mD)) = 0 \qquad (6.1)$$

by Serre duality.

It follows that for m sufficiently large, $H^2(X, \mathscr{O}_X(mC))$ vanishes, the Riemann–Roch theorem yields

$$h^0(X, \mathscr{O}_X(mC)) = m + \chi(X, \mathscr{O}_X) \geq 2$$

and the linear system $|mC|$ has no base-point (the only possible fixed curve is C, but $h^0(X, \mathscr{O}_X((m-1)C)) < h^0(X, \mathscr{O}_X(mC))$, and there are no isolated base-points since $C^2 = 0$). It defines a morphism from X to a projective space whose image is a curve. Its Stein factorization yields a morphism from X onto a smooth curve whose general fiber is numerically equivalent to C hence rational. All fibers are irreducible since $\mathbf{R}^+[C]$ is extremal.

Let us prove (d). Let D be a divisor on X with $D^2 > 0$ and $D \cdot H > 0$. For m sufficiently large, $H^2(X, \mathscr{O}_X(mD))$ vanishes by (6.1), and the Riemann–Roch theorem yields

$$h^0(X, \mathscr{O}_X(mD)) \geq \tfrac{1}{2}m^2 D^2 + O(m)$$

Since D^2 is positive, this proves that mD is linearly equivalent to an effective divisor for m sufficiently large, hence D is in NE(X). Therefore,

$$\{z \in N_1(X)_{\mathbf{R}} \mid z^2 > 0 \, , \ H \cdot z > 0\} \qquad (6.2)$$

is contained in NE(X). Since it is open, it is contained in its interior hence does not contain any extremal ray of $\overline{\mathrm{NE}}(X)$, except if X has Picard number 1. This proves (d).

Let us prove (e). Express r as the limit of a sequence of classes of effective divisors D_m. There exists an integer m_0 such that $r \cdot [D_{m_0}] < 0$; hence there exists an irreducible curve C such that $r \cdot C < 0$. Write

$$D_m = a_m C + D'_m$$

with $a_m \geq 0$ and D'_m effective with $C \cdot D'_m \geq 0$. Taking intersections with an ample divisor, we see that the upper limit of the sequence (a_m) is finite, so we may assume that it has a nonnegative limit a. In that case, $([D'_m])$ also has a limit $r' = r - a[C]$ in $\overline{\mathrm{NE}}(X)$ which satisfies

$$0 \leq r' \cdot C = r \cdot C - aC^2 < -aC^2$$

It follows that a is positive and C^2 is negative. Since $\mathbf{R}^+ r$ is extremal and $r = a[C] + r'$, the class r must be a multiple of $[C]$. $\square$

6.3. If K_X is nef, the effective cone lies entirely in the closed half-space $N_1(X)_{K_X \geq 0}$. In this case, the cone theorem does not give any information.

This is the case for example when X is an abelian surface. If we fix an ample divisor H on X, we have

$$\overline{NE}(X) = \{z \in N_1(X)_\mathbf{R} \mid z^2 \geq 0 \, , \, H \cdot z \geq 0\}$$

Indeed, one inclusion follows from the fact that any curve on X has non-negative self-intersection, and the other from (6.2). By Hodge theory, the intersection form on $N_1(X)_\mathbf{R}$ has exactly one positive eigenvalue, so that when this vector space has dimension 3, the closed cone of curves of X looks like this.

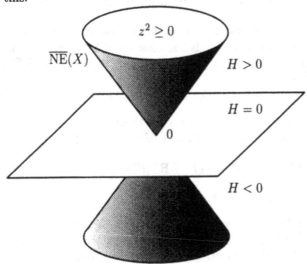

The effective cone of an abelian surface X.

In particular, it is not finitely generated. We will characterize in Exercise 6.7.6 all abelian varieties for which the closed cone of curves is finitely generated.

Every boundary point generates an extremal ray, hence there are extremal rays whose only rational point is 0: they cannot be generated by the class of a curve on X.

6.4. Let X be a $\mathbf{P}^1$-bundle over a smooth curve of genus g, so that $NE(X)$ is a convex cone in $\mathbf{R}^2$, hence has two extremal rays. The situation was discussed earlier in 1.35, using classical results from [H1], V, §2, but Lemma 6.2 allows us to recover most of the results. Let F be a fiber. Since $F^2 = 0$, its class lies in the boundary of $\overline{NE}(X)$ by Lemma 6.2(a), hence spans an extremal ray. Let ξ be the other extremal ray. Lemma 6.2(d) implies $\xi^2 \leq 0$.

- If $\xi^2 < 0$, we may, by Lemma 6.2(d), take for ξ the class of an irreducible curve C on X, and $NE(X) = \overline{NE}(X) = \mathbf{R}^+[C] + \mathbf{R}^+[F]$.

- If $\xi^2 = 0$, decompose ξ in a basis $([F], z)$ for $N_1(X)_{\mathbf{Q}}$ as $\xi = a[F] + bz$. Then $\xi^2 = 0$ implies that a/b is rational. However, it may happen that no multiple of ξ can be represented by an effective divisor, in which case $NE(X)$ is *not* closed (see 1.35).

Keeping the notation of 1.35, we can describe what the cone theorem yields in this situation. We have

$$K_X \cdot F = -2 \qquad \text{and} \qquad K_X \cdot C_0 = e + 2g - 2$$

When $e \geq 0$,

$$\overline{NE}(X) = \mathbf{R}^+[C_0] + \mathbf{R}^+[F] = \overline{NE}(X)_{K_X \geq 0} + \mathbf{R}^+[F]$$

except when $g = 0$ and $e = 0$ (in which case $X = \mathbf{P}^1 \times \mathbf{P}^1$) or $g = 0$ and $e = 1$ (in which case X is the blow-up of $\mathbf{P}^2$ at one point and C_0 is the exceptional divisor), in which case X is a Fano variety, $\overline{NE}(X)_{K_X \geq 0}$ is empty and

$$\overline{NE}(X) = \mathbf{R}^+[C_0] + \mathbf{R}^+[F]$$

is the decomposition given by the cone theorem.

When $e < 0$, we have $K_X \cdot (2C_0 + eF) = 4(g-1) \geq 0$ hence

$$\overline{NE}(X) = \mathbf{R}^+[2C_0 + eF] + \mathbf{R}^+[F] = \overline{NE}(X)_{K_X \geq 0} + \mathbf{R}^+[F]$$

Note that $(2C_0 + eF)^2 = 0$.

6.5. Let X be a Fano variety. The cone $\overline{NE}(X) - \{0\}$ is contained in the half-space $N_1(X)_{K_X < 0}$ (Theorem 1.27(a)), hence the set of extremal rays, being discrete and compact, is finite. The cone theorem yields

$$\overline{NE}(X) = NE(X) = \sum_{i=1}^{m} \mathbf{R}^+[\Gamma_i]$$

In particular, $NE(X)$ is closed. By Kleiman's criterion 1.27(a), a divisor on X is ample if and only if it has positive intersection with any curve.

Assume that X is a *surface* (then usually called a Del Pezzo surface). Since $-3 \leq K_X \cdot \Gamma_i < 0$, we have $-1 \leq \Gamma_i^2 \leq 1$ by the adjunction formula ([H1], V, ex. 1.3), and

- either $\Gamma_i^2 = 1$ for some i, the vector space $N_1(X)_{\mathbf{R}}$ has dimension 1 by Lemma 6.2(d) and X is isomorphic[1] to $\mathbf{P}^2$;

- or $\Gamma_i^2 = 0$ for some i, the surface X is ruled (Lemma 6.2(c)) and was studied in 6.4 (it is either $\mathbf{P}^1 \times \mathbf{P}^1$ or $\mathbf{P}^2$ blown up at a point);

---

[1]This is proved in [K1], th. III.3.7. The proof is easy over $\mathbf{C}$, but a bit more difficult in general.

- or the Γ_i are all exceptional curves.

When $N_1(X)_{\mathbf{R}}$ has dimension at least 3, we are in the last case. For example, when X is a smooth cubic surface,

$$\mathrm{NE}(X) = \sum_{i=1}^{27} \mathbf{R}^+[\Gamma_i] \subset \mathbf{R}^7$$

where the Γ_i are the 27 lines on X.

6.6. Let $X \to \mathbf{P}^2$ be the blow-up at the nine base-points of a general pencil of cubics, let $\pi : X \to \mathbf{P}^1$ be the morphism given by the pencil of cubics, and let B be the finite subset of X where π is not smooth. The exceptional divisors $E_0, \ldots, E_8$ are sections of π. Smooth fibers of π are elliptic curves, hence become abelian groups by choosing their intersection with E_0 as the origin. Translations by elements of E_i then generate a subgroup of $\mathrm{Aut}(X - B)$ which can be shown to be isomorphic to $\mathbf{Z}^8$.

Since X is smooth, any automorphism σ of $X - B$ extends to X Indeed, if $\tilde{\sigma} : \tilde{X} \xrightarrow{\varepsilon} X \xdashrightarrow{\sigma} X$ is a resolution of its indeterminacies and E the last (-1)-curve of the composition of blow-ups ε, its image in X is a curve, and any point x of $\tilde{\sigma}(E) - B$ has at least two preimages in $\tilde{X}$ (one on E, and $\varepsilon^{-1}(\sigma^{-1}(x))$). By Zariski's theorem, the fibers of $\tilde{\sigma}$ are connected, hence positive-dimensional above $\tilde{\sigma}(E)$, which is absurd. It follows that there are no exceptional curves in $\tilde{X}$, hence ε is an isomorphism.

For any such σ, the curve $E_\sigma = \sigma(E_0)$ is rational with self-intersection -1 and $K_X \cdot E_\sigma = -1$ (exceptional curves are preserved by deformation, so they are still there when the 9 points are moved in general position; see [H1], V, ex. 4.15(e), for a direct construction).

It follows from Lemma 6.2(b) that $\overline{\mathrm{NE}}(X)$ has infinitely many extremal rays contained in the open half-space $N_1(X)_{K_X < 0}$, which are *not* locally finite in a neighborhood of $K_X^\perp$, because $K_X \cdot E_\sigma = -1$ but $(E_\sigma)_{\sigma \in \mathbf{Z}^8}$ is unbounded since the set of classes of irreducible curves is discrete in $N_1(X)_{\mathbf{R}}$.

6.3   Elementary Properties of Cones

Let V be a cone in $\mathbf{R}^m$. We define its dual cone by

$$V^* = \{\ell \in (\mathbf{R}^m)^* \mid \ell \geq 0 \text{ on } V\}$$

Recall that a subcone W of V is *extremal* if it is closed and convex and if any two elements of V whose sum is in W are both in W. An extremal subcone of dimension 1 is called an *extremal ray*. A nonzero linear form ℓ in V^* is a *supporting function* of the extremal subcone W if it vanishes on W.

Lemma 6.7 *Let V be a closed convex cone in $\mathbf{R}^m$.*

(a) *We have $V = V^{**}$ and*

$$V \text{ contains no lines} \iff V^* \text{ spans } (\mathbf{R}^m)^*$$

The interior of V^* is

$$\{\ell \in (\mathbf{R}^m)^* \mid \ell > 0 \text{ on } V - \{0\}\}$$

(b) *If V contains no lines, it is the convex hull of its extremal rays.*

(c) *Any proper extremal subcone of V has a supporting function.*

(d) *If V contains no lines [2] and W is a proper closed subcone of V, there exists a linear form in V^* that is positive on $W - \{0\}$ and vanishes on some extremal ray of V.*

PROOF. Obviously, V is contained in V^{**}. Choose a scalar product on $\mathbf{R}^m$. If $z \notin V$, let $p_V(z)$ be the projection of z on the closed convex set V. Since V is a cone, $z - p_V(z)$ is orthogonal to $p_V(z)$. The linear form $\langle p_V(z) - z, \cdot \rangle$ is nonnegative on V and negative at z, hence $z \notin V^{**}$.

If V contains a line L, any element of V^* must be nonnegative, hence must vanish, on L: the cone V^* is contained in $L^\perp$. Conversely, if V^* is contained in a hyperplane H, its dual V contains the line $H^\perp$ in $\mathbf{R}^m$.

Let ℓ be an interior point of V^*. For any nonzero z in V, there exists a linear form ℓ' with $\ell'(z) > 0$ and small enough so that $\ell - \ell'$ is still in V^*. This implies $(\ell - \ell')(z) \geq 0$, hence $\ell(z) > 0$. Since the set

$$\{\ell \in (\mathbf{R}^m)^* \mid \ell > 0 \text{ on } V - \{0\}\}$$

is open, this proves (a).

Assume that V contains no lines. We will prove by induction on m that any point of V is in the linear span of m extremal rays.

6.8. Note that for any point v of ∂V, there exists by (a) a nonzero element ℓ in V^* that vanishes at v. An extremal ray $\mathbf{R}^+ r$ in $\mathrm{Ker}(\ell) \cap V$ (which exists thanks to the induction hypothesis) is still extremal in V: if $r = x_1 + x_2$ with x_1 and x_2 in V, since $\ell(x_i) \geq 0$ and $\ell(r) = 0$, we get $x_i \in \mathrm{Ker}(\ell) \cap V$, hence they are both proportional to r.

Given $v \in V$, the set $\{\lambda \in \mathbf{R}^+ \mid v - \lambda r \in V\}$ is a closed nonempty interval which is bounded above (otherwise $-r = \lim_{\lambda \to +\infty} \frac{1}{\lambda}(v - \lambda r)$ would be in V). If λ_0 is its maximum, $v - \lambda_0 r$ is in ∂V, hence there exists by (a) an element ℓ' of V^* that vanishes at $v - \lambda_0 r$. Since

$$v = \lambda_0 r + (v - \lambda_0 r)$$

---

[2] This assumption is necessary, as shown by the example $V = \{(x, y) \in \mathbf{R}^2 \mid y \geq 0\}$ and $W = \{(x, y) \in \mathbf{R}^2 \mid x, y \geq 0\}$.

item (b) follows from the induction hypothesis applied to the closed convex cone $\text{Ker}(\ell') \cap V$ and the fact that any extremal ray in $\text{Ker}(\ell') \cap V$ is still extremal for V.

Let us prove (c). We may assume that V spans $\mathbf{R}^m$. Note that an extremal subcone W of V distinct from V is contained in ∂V: if W contains an interior point v, then for any small x, we have $v \pm x \in V$ and $2v = (v + x) + (v - x)$ implies $v \pm x \in W$. Hence W is open in the interior of V. Since it is closed, it contains it.

In particular, the interior of W is empty, hence its span $\langle W \rangle$ is not $\mathbf{R}^m$. Let w be a point of the interior of W in $\langle W \rangle$. By (a), there exists a nonzero element ℓ of V^* that vanishes at w. By (a) again (applied to W^* in its span), ℓ must vanish on $\langle W \rangle$, hence is a supporting function of W.

Let us prove (d). Since W contains no lines, there exists by (a) a point in the interior of W^* which is not in V^*. The segment connecting it to a point in the interior of V^* crosses the boundary of V^* at a point in the interior of W^*. This point corresponds to a linear form ℓ that is positive on $W - \{0\}$ and vanishes at a nonzero point of V. By (b), the closed cone $\text{Ker}(\ell) \cap V$ has an extremal ray, which is still extremal in V by 6.8. This proves (d). $\qquad\qquad\qquad\qquad\qquad\qquad\qquad\qquad\qquad\qquad$ $\square$

6.4 Proof of the Cone Theorem

We prove the cone theorem: if X is a smooth projective variety,

$$\overline{\text{NE}}(X) = \overline{\text{NE}}(X)_{K_X \geq 0} + \sum_{i \in I} \mathbf{R}^+ [\Gamma_i]$$

where Γ_i are rational curves on X such that $0 < -K_X \cdot \Gamma_i \leq \dim(X) + 1$.

The idea is quite simple: if $\overline{\text{NE}}(X)$ is not equal to the closure of the right-hand side, there exists a divisor M on X which is nonnegative on $\overline{\text{NE}}(X)$ (hence nef), positive on the closure of the right-hand side, and vanishes at some nonzero point z of $\overline{\text{NE}}(X)$, which must therefore satisfy $K_X \cdot z < 0$. We approximate M by an ample divisor, z by an effective 1-cycle and use the bend-and-break theorem 3.6 to get a contradiction. In the third and last step, we prove that the right-hand side is closed by a formal argument with no geometric content.

PROOF OF THE THEOREM. As we saw in Chapter 2, there are only countably many families of, hence classes of, rational curves on X. Pick a representative Γ_i for each such class z_i that satisfies $0 < -K_X \cdot z_i \leq \dim(X) + 1$.

First step: the rays $\mathbf{R}^+ z_i$ are locally discrete in the half-space $N_1(X)_{K_X < 0}$.
Let H be an ample divisor on X. It is enough to show that for each $\varepsilon > 0$, there are only finitely many classes z_i in the half-space $N_1(X)_{K_X + \varepsilon H < 0}$,

since the union of these half-spaces is $N_1(X)_{K_X<0}$. If $(K_X + \varepsilon H) \cdot \Gamma_i < 0$, we have

$$H \cdot \Gamma_i < \varepsilon^{-1} K_X \cdot \Gamma_i \le \varepsilon^{-1}(\dim(X) + 1)$$

and there are finitely many such classes of curves on X (Theorem 1.27(b)).

Second step: $\overline{\mathrm{NE}}(X)$ is equal to the closure of

$$V = \overline{\mathrm{NE}}(X)_{K_X \ge 0} + \sum_i \mathbf{R}^+ z_i$$

If this is not the case, there exists by Lemma 6.7(d) (since $\overline{\mathrm{NE}}(X)$ contains no lines) an $\mathbf{R}$-divisor M on X that is nonnegative on $\overline{\mathrm{NE}}(X)$ (it is in particular nef), positive on $\overline{V} - \{0\}$, and that vanishes at some nonzero point z of $\overline{\mathrm{NE}}(X)$. This point cannot be in V, hence $K_X \cdot z < 0$.

Choose a norm on $N_1(X)_{\mathbf{R}}$ such that $\|[C]\| \ge 1$ for each irreducible curve C (this is possible since the set of classes of irreducible curves is discrete). We may assume, upon replacing M with a multiple, that $M \cdot v \ge 2\|v\|$ for all v in $\overline{V}$. Since the class $[M]$ is a limit of classes of ample $\mathbf{Q}$-divisors, and z is a limit of classes of effective rational 1-cycles, there exist an ample $\mathbf{Q}$-divisor H and an effective 1-cycle Z such that

$$2\dim(X)(H \cdot Z) < -K_X \cdot Z \quad \text{and} \quad H \cdot v \ge \|v\| \qquad (6.3)$$

for all v in $\overline{V}$. We may further assume, by throwing away the other components, that each component C of Z satisfies $-K_X \cdot C > 0$.

Since the class of every rational curve Γ on X such that $-K_X \cdot \Gamma \le \dim(X) + 1$ is in $\overline{V}$, we have $H \cdot \Gamma \ge 1$ by (6.3). Since X is smooth, the bend-and-break theorem 3.6 implies

$$2\dim(X) \frac{H \cdot C}{-K_X \cdot C} \ge H \cdot \Gamma \ge 1$$

for every component C of Z. This contradicts the first inequality in (6.3) and finishes the proof of the second step.

Third step: for any set J of indices, the cone

$$\overline{\mathrm{NE}}(X)_{K_X \ge 0} + \sum_{j \in J} \mathbf{R}^+ z_j$$

is closed.

Let V_J be this cone. By Lemma 6.7(b), it is enough to show that any extremal ray $\mathbf{R}^+ r$ in $\overline{V_J}$ satisfying $K_X \cdot r < 0$ is in V_J. Write r as the limit of a sequence $(r_m + s_m)$, where $K_X \cdot r_m \ge 0$ and s_m is in $\sum_{j \in J} \mathbf{R}^+ z_j$. Let H be an ample divisor on X. The sequences $(H \cdot r_m)$ and $(H \cdot s_m)$ are bounded (by $H \cdot r + 1$ for m large) hence, by Theorem 1.27(b), we may assume after taking subsequences that both sequences (r_m) and (s_m) have limits, which are in $\overline{V_J}$. Because r spans an extremal ray in $\overline{V_J}$, the limits must be nonnegative multiples of r. Since $K_X \cdot r < 0$, the limit of (r_m) must vanish. Let ε be a positive number such that $(K_X + \varepsilon H) \cdot r < 0$. By the first step, there are only finitely many classes $z_{j_1}, \ldots, z_{j_q}$, with $j_\alpha \in J$, such that $(K_X + \varepsilon H) \cdot z_{j_\alpha} < 0$. Write s_m as $s'_m + s''_m$, where

$$s'_m = \sum_{\alpha=1}^{q} \lambda_{\alpha,m} z_{j_\alpha}$$

and $(K_X + \varepsilon H) \cdot s''_m \geq 0$. As above, we may assume after taking subsequences that the sequences (s'_m) and $(\lambda_{\alpha,m} z_{j_\alpha})_m$ all converge to nonnegative multiples of r. Since $(K_X + \varepsilon H) \cdot r < 0$, the limit of (s''_m) must vanish; r is a multiple of one the z_{j_α}, hence is in V_J.

The proof shows that any extremal ray of $\overline{NE}(X)$ is proportional to a z_i, hence is in V_I, where I is the set of indices i such that $\mathbf{R}^+ z_i$ is an extremal ray in $\overline{NE}(X)$. This finishes the proof of the cone theorem.   □

Corollary 6.9 *Let X be a smooth projective variety and let R be a K_X-negative extremal ray. There exists a nef divisor M_R on X such that*

(a) $R = \{z \in \overline{NE}(X) \mid M_R \cdot z = 0\}$;

(b) *the divisor $m M_R - K_X$ is ample for all integers m sufficiently large.*

The divisor M_R will be called a *supporting divisor* for R. Property (b) is useful in conjunction with Theorem 7.32: it implies that in characteristic zero, the linear system $|m M_R|$ is base-point-free for all integers m sufficiently large, and defines the contraction of R.

PROOF OF THE COROLLARY. With the notation of the proof of the cone theorem, there exists a (unique) element i_0 of I such that $R = \mathbf{R}^+ z_{i_0}$. By the third step of the proof, the subcone

$$V = V_{I - \{i_0\}} = \overline{NE}(X)_{K_X \geq 0} + \sum_{i \in I, i \neq i_0} \mathbf{R}^+ z_i$$

of $\overline{NE}(X)$ is closed and distinct from $\overline{NE}(X)$ since it does not contain R. By Lemma 6.7(d), there exists a linear form that is nonnegative on $\overline{NE}(X)$, positive on $V - \{0\}$, and that vanishes at some nonzero point of $\overline{NE}(X)$, hence on R since $\overline{NE}(X) = V + R$. The intersection of the interior of V^* and the *rational* hyperplane $R^\perp$ is therefore nonempty, hence contains an integral point: there exists a divisor M_R on X that is positive on $V - \{0\}$ and vanishes on R. It is in particular nef and (a) holds.

Choose a norm on $N_1(X)_{\mathbf{R}}$ and let a be the (positive) minimum of M_R on the set of elements of V with norm 1. If b is the maximum of K_X on the same compact, the divisor $m M_R - K_X$ is positive on $V - \{0\}$ for m rational greater than b/a, and positive on $R - \{0\}$ for $m \geq 0$, hence ample for $m > \max(b/a, 0)$ by Kleiman's criterion 1.27(a). This proves (b).   □

6.5   Contractions of Extremal Rays

As we saw in Section 1.11, varieties with nef canonical divisor have a simple birational structure. If the canonical divisor is not nef, there exists by the cone theorem a K_X-negative extremal ray. The central idea of Mori's program is to contract this extremal ray to get hopefully a simpler variety, and to repeat the process to end up with a variety with nef canonical divisor. Unfortunately, things are not that simple.

The contraction of a K_X-negative extremal ray R on a smooth variety X does exist in characteristic zero (with the notation of Corollary 6.9, it is the Stein factorization of the morphism associated with some base-point-free multiple of M_R; the proof will be given in Theorem 7.39) and in any characteristic when X is a surface (see Exercise 6.7.1). By Proposition 1.14(b), it is uniquely defined and will be denoted by c_R. It contracts all curves whose class lies in R (with the terminology of Section 1.1, the *relative cone* of the contraction is therefore R). The union of all these curves is called the *locus* of R and will be denoted by $\mathrm{locus}(R)$.

We now describe geometric properties of contractions.

Proposition 6.10 *Let X be a smooth projective variety and let R be a K_X-negative extremal ray of $\overline{\mathrm{NE}}(X)$. The locus of R is closed. If Z is an irreducible component of R,*

(a) *the variety Z is uniruled;*

(b) *if Z has codimension 1, it is equal to the locus of R;*

(c) *if the contraction of R exists, the following inequality holds:*

$$\dim(Z) \geq \tfrac{1}{2}(\dim(X) + \dim(c_R(Z)))$$

The locus of R may be disconnected (see 6.19). The inequality in (c) is sharp (see 6.17), but can be made more precise (see 6.11).

PROOF OF THE PROPOSITION. Any point x in the locus of R is on some irreducible curve C whose class is in R. Let M_R be a nef divisor on X as in Corollary 6.9, let H be an ample divisor on X and let m be an integer such that

$$m > 2\dim(X)\frac{H \cdot C}{-K_X \cdot C}$$

(note that this number only depends on R, not on C). By Theorem 3.6, applied with the ample divisor $mM_R + H$, there exists a rational curve Γ

through x such that

$$
\begin{aligned}
0 \;&<\; (mM_R + H) \cdot \Gamma \\
&\leq\; 2\dim(X)\frac{(mM_R + H) \cdot C}{-K_X \cdot C} \\
&=\; 2\dim(X)\frac{H \cdot C}{-K_X \cdot C} \\
&<\; m
\end{aligned}
$$

from which it follows $M_R \cdot \Gamma = 0$ and $H \cdot \Gamma < m$: the class $[\Gamma]$ is in R hence Γ is contained in $\operatorname{locus}(R)$.

The locus of R is therefore the union of all rational curves of H-degree at most m whose class is in R. The proof of Lemma 3.7 shows that any point in the closure of $\operatorname{locus}(R)$ is on some effective rational 1-cycle $\sum_i a_i\Gamma_i$ whose class is in R. Since R is extremal, each $[\Gamma_i]$ must be in R and x is in $\operatorname{locus}(R)$. This proves that $\operatorname{locus}(R)$ is closed, and (a). The same proof shows that the locus of any extremal subcone of $\overline{\mathrm{NE}}(X)$, similarly defined, is also closed in X. Of course, once we know that the contraction exists, $\operatorname{locus}(R)$ is closed since it is the set of points where the contraction is not an isomorphism.

Assume that the locus of R is not X. It can be proved directly that $M_R^{\dim(X)}$ is positive ([K1], th. III.1.6). Since it also follows if we assume the existence of a contraction given by some base-point-free multiple of M_R (which will be proved in Theorem 7.39 in characteristic zero), we will not prove it. By Proposition 1.31, there is an effective divisor D linearly equivalent to $mM_R - H$ for m sufficiently large. A nonzero element in R has negative intersection with D, hence with some irreducible component D' of D. Any irreducible curve with class in R must then be contained in D', which therefore contains the locus of R. This implies (b).

Assume now that x is general in Z and pick a rational curve Γ in Z through x with class in R and minimal $(-K_X)$-degree (note that the class of any such curve is in R, hence has positive $(-K_X)$-degree). Let $f : \mathbf{P}^1 \to \Gamma \subset X$ be the normalization, with $f(0) = x$.

Let T be a component of $\operatorname{Mor}(\mathbf{P}^1, X)$ passing through $[f]$ and let $e_0 : T \to X$ be the map $t \mapsto f_t(0)$. By (2.2), T has dimension at least $\dim(X) + 1$. Each curve $f_t(\mathbf{P}^1)$ has same class as Γ, hence is contained in Z. In particular, for any component T_x of $e_0^{-1}(x)$, we have

$$
\begin{aligned}
\dim(Z) \;&\geq\; \dim(T) - \dim(T_x) \\
&\geq\; \dim(X) + 1 - \dim(T_x) \tag{6.4}
\end{aligned}
$$

Consider the evaluation $e_\infty : T_x \to X$ and let y be a point in X distinct from x. If $e_\infty^{-1}(y)$ *has dimension at least* 2, Proposition 3.2 implies that Γ is numerically equivalent to a connected effective rational nonintegral 1-cycle $\sum_i a_i\Gamma_i$ passing through x and y. Since R is extremal, each $[\Gamma_i]$ must be in

R, hence $0 < -K_X \cdot \Gamma_i < -K_X \cdot \Gamma$ for each i. This contradicts the choice of Γ.

It follows that a general fiber of e_∞ has dimension at most 1. Since the curve $f_t(\mathbf{P}^1)$, for $t \in T_x$, passes through x hence has same image as x by c_R,

$$e_\infty(T_x) = \bigcup_{t \in T_x} \{f_t(\infty)\} = \bigcup_{t \in T_x} f_t(\mathbf{P}^1)$$

is irreducible and contained in the fiber $c_R^{-1}(c_R(x))$. We get

$$\dim_x(c_R^{-1}(c_R(x))) \geq \dim(\overline{e_\infty(T_x)}) \geq \dim(T_x) - 1 \qquad (6.5)$$

Since the left-hand side is $\dim(Z) - \dim(c_R(Z))$, item (c) follows from (6.4). □

6.11. Length of an extremal ray. Inequality (2.2) actually yields

$$\dim(Z) \geq \dim(X) - K_X \cdot \Gamma - \dim(T_x)$$

hence the following improvement of (6.5): any positive-dimensional irreducible component F of a fiber of c_R satisfies

$$\dim(F) \geq \operatorname{codim}(Z) - K_X \cdot \Gamma - 1$$

In particular, F is covered by rational curves of $(-K_X)$-degree at most $\dim(F) + 1 - \operatorname{codim}(\operatorname{locus}(R))$. The integer

$$\ell(R) = \min\{-K_X \cdot \Gamma \mid \Gamma \text{ rational curve on } X \text{ with class in } R\}$$

is called the *length* of the extremal ray R. The inequality

$$\dim(F) \geq \operatorname{codim}(\operatorname{locus}(R)) + \ell(R) - 1 \qquad (6.6)$$

valid for any positive-dimensional irreducible component F of a fiber of c_R, is due to Wiśniewski ([W]).

6.12. Types of contractions. There are three cases:

- the locus of R is X (the contraction, if it exists, is called a *fiber contraction*);

- the locus of R is a divisor, which is irreducible by Proposition 6.10 (the contraction, if it exists, is called a *divisorial contraction*);

- the locus of R has codimension at least 2 (the contraction, if it exists, is called a *small contraction*).

Assume in what follows that X is a smooth projective variety and that R is a K_X-negative extremal ray whose contraction c_R exists.

In the first case, X is uniruled. The image of the contraction has dimension less than X and its general fiber is a Fano variety (7.42). An example is given in 6.14 below.

In the second case, the image of the contraction may be singular, but not too much: some multiple of the canonical divisor is a Cartier divisor.[3] An example is given in 6.15 below. *This shows that in order to be able to continue Mori's program, we must allow singularities.*

The last case causes a lot of trouble: by footnote 10, p. 28, the image of the contraction is not locally factorial. Actually, no multiple of the canonical divisor is a Cartier divisor, so it does not even make sense to say that it is nef! By Proposition 6.10 (or 6.11), the fibers of c_R contained in locus(R) have dimension at least 2 and

$$\dim(X) \geq \dim(c_R(\text{locus}(R))) + 4$$

In particular, there are no small extremal contractions for smooth varieties in dimension 3 (see 6.20 for an example with mildly singular varieties).

6.13. In this case, it is impossible to continue the program with $c_R(X)$. Instead, Mori's idea is that there should exist another (mildly singular) projective variety X^+ with a small contraction $c^+ : X^+ \to c_R(X)$ such that K_{X^+} has *positive* degree on curves contracted by c^+. The map c^+ (or sometimes the resulting rational map $(c^+)^{-1} \circ c : X \dashrightarrow X^+$) is called a *flip* (see Definition 7.43 for more details). An example is given in 6.17 below.

6.6 Examples of Contractions of Extremal Rays

We give a list of examples of contractions of extremal rays and flips of small contractions. The situation can be quite subtle (see, for example, 6.18).

6.14. A projective bundle is a fiber contraction. Let $\mathscr{E}$ be a vector bundle of rank r over a smooth projective variety Y and let $X = \mathbf{P}(\mathscr{E})$,[4] with projection $\pi : X \to Y$. If ξ is the class of the line bundle $\mathscr{O}_X(1)$, we have

$$K_X = -(r+1)\xi + \pi^*(K_Y + \det(\mathscr{E}))$$

If ℓ is the class of a line contained in a fiber of π, we have $K_X \cdot \ell = -(r+1)$. The class ℓ spans a K_X-negative ray whose contraction is π: indeed, a curve

---

[3]For any projective variety X that is nonsingular in codimension 1, we define a canonical (Weil) divisor class K_X as the class of the closure of a canonical divisor of the smooth locus of X (see [H1], II, prop. 6.5(b)).

[4]As usual, we follow Grothendieck's notation: for a vector bundle $\mathscr{E}$, the projectivization $\mathbf{P}(\mathscr{E})$ is the space of *hyperplanes* in the fibers of $\mathscr{E}$.

is contracted by π if and only if it is numerically equivalent to a multiple of ℓ (by Proposition 1.14(a), this implies that the ray spanned by ℓ is extremal).

6.15. A smooth blow-up is a divisorial contraction. Let Y be a smooth projective variety, let Z be a smooth subvariety of Y of codimension c, and let $\pi : X \to Y$ be the blow-up of Z, with exceptional divisor E. We have ([H1], II, ex. 8.5(b))

$$K_X = \pi^* K_Y + (c-1)E$$

Any fiber F of $E \to Z$ is isomorphic to $\mathbf{P}^{c-1}$, and $\mathscr{O}_F(E)$ is isomorphic to $\mathscr{O}_F(-1)$. If ℓ is the class of a line contained in F, we have $K_X \cdot \ell = -(c-1)$. The class ℓ therefore spans a K_X-negative ray whose contraction is π: a curve is contracted by π if and only if it lies in a fiber of $E \to Z$, hence is numerically equivalent to a multiple of ℓ.

6.16. A fiber contraction that is not a projective bundle, and a divisorial contraction that is not a smooth blow-up. Let C be a smooth curve of genus g, let d be a positive integer, and let $J^d(C)$ be the Jacobian of C which parametrizes isomorphism classes of invertible sheaves of degree d on C.

Let C_d be the symmetric product of d copies of C and let $\pi_d : C_d \to J^d(C)$ be the Abel–Jacobi map. For $d > 2g - 1$, it is a $\mathbf{P}^{d-g}$-bundle hence the contraction of a K_{C_d}-negative extremal ray by 6.14.

For any positive d, all fibers of π_d are projective spaces. If ℓ_d is the class of a line in a fiber, we have

$$K_{C_d} \cdot \ell_d = g - d - 1$$

Indeed, the formula holds for $d > 2g - 1$ by 6.14. Assume it holds for d. Pick a point of C to get an embedding $\iota : C_{d-1} \to C_d$. Then $\iota^* C_{d-1} \cdot \ell_d = 1$ and the adjunction formula yields

$$
\begin{aligned}
K_{C_{d-1}} \cdot \ell_{d-1} &= \iota^*(K_{C_d} + C_{d-1}) \cdot \ell_{d-1} \\
&= (K_{C_d} + C_{d-1}) \cdot \iota_* \ell_{d-1} \\
&= (K_{C_d} + C_{d-1}) \cdot \ell_d \\
&= (g - d - 1) + 1
\end{aligned}
$$

which proves the formula by descending induction on d.

It follows that for $d \geq g$, the (surjective) map π_d is the contraction of the K_{C_d}-negative extremal ray $\mathbf{R}^+ \ell_d$. It is a fiber contraction for $d > g$, a divisorial contraction for $d = g$. For $d = g + 1$, the general fiber is $\mathbf{P}^1$, but there are bigger fibers when $g \geq 3$, so the contraction is not a projective bundle. For $d = g$, the contraction of the locus of $\mathbf{R}^+ \ell_d$ is

$$\{L \in J^g(C) \mid h^0(C, \omega_C \otimes L^{-1}) > 0\}$$

and the general fiber is a $\mathbf{P}^1$, but there are bigger fibers when $g \geq 6$, because the curve C has a g^1_{g-2}, and the contraction is not a smooth blow-up.

6.17. A small contraction and a flip. Let r and s be positive integers, and let $X_{r \cdot s}$ and $Y_{r,s}$ be the varieties defined in 1.36 (see also Example 3.16(2)). The variety $X_{r \cdot s}$ is a Fano variety,

$$\mathrm{NE}(X_{r \cdot s}) = \overline{\mathrm{NE}}(X_{r \cdot s}) = R + R' + R''$$

(see (1.10)) and the rays R, R', and R'' are all $K_{X_{r \cdot s}}$-negative. The corresponding contractions fit into the diagram

$$
\begin{array}{ccccc}
Y_{s,r} & \xleftarrow{\;c_R\;} & X_{r \cdot s} & \xrightarrow{\;c_{R''}\;} & Y_{r,s} \\
\downarrow{\scriptstyle c_{R'}} & & \downarrow{\scriptstyle c_{R'}} & & \downarrow{\scriptstyle c_{R'}} \\
\mathbf{P}^r & \xleftarrow{\;c_R\;} & \mathbf{P}^r \times \mathbf{P}^s & \xrightarrow{\;c_{R''}\;} & \mathbf{P}^s
\end{array}
$$

The contractions corresponding to the three extremal $K_{X_{r \cdot s}}$-negative planes in $\mathrm{NE}(X_{r \cdot s})$ are

- for $\langle R, R' \rangle$, the map $c_R \circ c_{R'}$ to $\mathbf{P}^r$;

- for $\langle R', R'' \rangle$, the map $c_{R''} \circ c_{R'}$ to $\mathbf{P}^s$;

- for $\langle R, R'' \rangle$, a map $X_{r \cdot s} \to \widehat{Y}_{r \cdot s}$ which, by Proposition 1.14(b), must factor through both c_R and $c_{R''}$.

Assume from now on $r < s$. The variety $Y_{r,s}$ is a Fano variety (Example 3.3(2)) and the contraction $c_R : Y_{r,s} \to \widehat{Y}_{r \cdot s}$ of the $K_{Y_{r,s}}$-negative extremal ray $R = \mathbf{R}^+\ell$ is the morphism associated with the base-point-free linear system $|\mathcal{O}_{Y_{r,s}}(1)|$. The locus of R is the variety $P_{r,s}$ defined in 1.36, which has dimension s and is contracted to a point; c_R is therefore a small contraction, and when $r = s - 1$, there is equality in Proposition 6.10(c).

The corresponding flip is the composition $Y_{r,s} \xdashrightarrow{c_{R''}^{-1}} X_{r \cdot s} \xrightarrow{c_R} Y_{s,r}$. It is an isomorphism in codimension 1 (more exactly, outside the subvarieties $P_{r,s}$ and $P_{s,r}$), so the Picard numbers of $Y_{r,s}$ and $Y_{s,r}$ are the same, but the $K_{Y_{r,s}}$-negative extremal ray R on $Y_{r,s}$ has been "replaced" by the $K_{Y_{s,r}}$-*positive* extremal ray R'' on $Y_{s,r}$.

6.18. A divisorial contraction with singular image. Let Z be a smooth projective threefold and let C be an irreducible curve in Z whose only singularity is a node. The blow-up Y of Z along C is normal and its only singularity is an ordinary double point q. This is checked by a local calculation: locally analytically, the ideal of C is generated by xy and z, where x, y, z form a system of parameters. The blow-up is

$$\{((x,y,z),[u,v]) \in \mathbf{A}^3 \times \mathbf{P}^1 \mid xyv = zu\}$$

It is smooth except at the point $q = ((0,0,0),[0,1])$. The exceptional divisor is the $\mathbf{P}^1$-bundle over C with local equations $xy = z = 0$.

The blow-up X of Y at q is smooth, with exceptional divisor a smooth quadric Q. The proper transform E of the exceptional divisor of Y meets Q in the union of two lines L_1 and L_2 belonging to the two different rulings of Q. Let $\tilde{E} \to E$ and $\tilde{C} \to C$ be the normalizations. Each fiber of $\tilde{E} \to \tilde{C}$ is a smooth rational curve, except over the two preimages p_i of the node of C, where it is the union $L_{i1} \cup L_{i2}$ of two smooth rational curves meeting transversally. For $i = 1, 2$, the curve L_{i1} maps onto L_i, the other curve L_{i2} onto the same rational curve L. *It follows that $L + L_1$ and $L + L_2$ are algebraically equivalent on X, hence L_1 and L_2 have the same class ℓ.*

Any curve contracted by the blow-up $\pi : X \to Y$ is contained in Q, hence its class is a multiple of ℓ. A local calculation[5] shows that $\mathscr{O}_Q(K_X)$ is of type $(-1, -1)$, hence $K_X \cdot \ell = -1$. The ray $\mathbf{R}^+\ell$ is K_X-negative and its (divisorial) contraction is π (hence $\mathbf{R}^+\ell$ is extremal).

This situation is very subtle: although the completion of the local ring $\mathscr{O}_{Y,q}$ is not factorial,[6] the fact that L_1 is numerically equivalent to L_2 implies that the ring $\mathscr{O}_{Y,q}$ is factorial (see [Mo2], (3.31)). This phenomenon also occurs in the following situation: let Y be a hypersurface of degree at least 3 in $\mathbf{P}^4$ whose only singularity is an ordinary double point q. It is shown in [G], §15, that in the (smooth) blow-up X of Y at q, the two lines of the ruling of the exceptional divisor are again numerically equivalent (although they are not algebraically equivalent in general by [CC]). The blow-up $X \to Y$ is again the contraction of the K_X-negative extremal ray generated by the class of these lines.

Assume more generally that a smooth projective threefold X contains a divisor Q isomorphic to $\mathbf{P}^1 \times \mathbf{P}^1$, and let L_1 and L_2 be lines in each ruling of Q. Assume that $K_X \cdot L_i$ and $Q \cdot L_i$ are negative for $i = 1$ or 2 (by adjunction, the only possible case is $K_X \cdot L_i = Q \cdot L_i = -1$). Proposition 6.10(b) implies that the locus of each ray $R_i = \mathbf{R}^+[L_i]$ is Q. There are two cases:

- either L_1 and L_2 are numerically equivalent (this is the case discussed above). The K_X-negative ray $R_1 = R_2$ is extremal and its contraction, if it exists, yields a singular, but locally factorial, variety ([Mo2], th. (3.3) and cor. (3.4));

- or L_1 and L_2 are *not* numerically equivalent. At least one the rays R_1 or R_2 is extremal, and its contraction, if it exists, is the blow-up of a smooth variety along a smooth rational curve with normal bundle $\mathscr{O}(-1) \oplus \mathscr{O}(-1)$.[7]

---

[5]Analytically, a neighborhood of the double point is given by an equation $xy = zu$ in $\mathbf{A}^4$. Its blow-up is the total space of the $\mathbf{P}^1$-bundle $\mathbf{P}(\mathscr{O}_Q \oplus \mathscr{O}_Q(1)) \to Q$, where Q is a smooth quadric in $\mathbf{P}^3$, and the exceptional divisor is the image of the section corresponding to the trivial quotient of $\mathscr{O}_Q \oplus \mathscr{O}_Q(1)$. Its normal bundle is $\mathscr{O}_Q(1)^*$.

[6]It is isomorphic to $\mathbf{k}[[x, y, z, u]]/(xy - zu)$, and the equality $xy = zu$ is a decomposition in a product of irreducibles in two different ways; see also [H1], II, ex. 6.5(b).

[7]This can be seen as follows: any irreducible curve C such that $Q \cdot C$ is negative is contained in Q, hence is numerically equivalent to a linear combination of L_1 and

In both cases, $\mathcal{O}_Q(Q)$ is of type $(-1,-1)$. It is impossible to tell, locally analytically, the difference between the two cases.

6.19. A small contraction with disconnected exceptional locus

This construction is due to Kawamata ([Ka5], p. 599). Start from a smooth complex fourfold X'' that contains a smooth curve C'' and a smooth surface S'' meeting transversely at points $x_1, \ldots, x_r$. Let $\varepsilon' : X' \to X''$ be the blow-up of C''. The exceptional divisor C' is a smooth threefold which is a $\mathbf{P}^2$-bundle over C''. The strict transform S' of S'' is the blow-up of S'' at the points $x_1, \ldots, x_r$. Let $E'_1, \ldots, E'_r$ be the corresponding exceptional curves and let $P'_1, \ldots, P'_r$ be the corresponding $\mathbf{P}^2$ that contain them, i.e., $P'_i = \varepsilon'^{-1}(x_i)$. Let $\varepsilon : X \to X'$ be the blow-up of S'. The exceptional divisor S is a smooth threefold which is a $\mathbf{P}^1$-bundle over S'. Let Γ_i be the fiber over a point of E'_i and let P_i be the strict transform of P'_i. Finally, let L be a line in one of the $\mathbf{P}^2$ in the inverse image C of C'.

For $r = 1$, the picture is something like the following diagram.

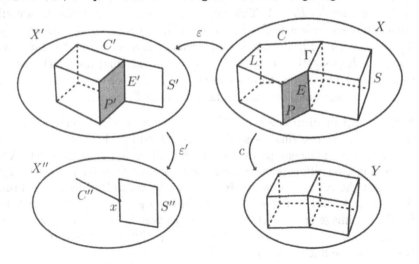

A small contraction.

L_2. It follows that any element of $\overline{NE}(X)$ can be written as $a_1[L_1] + a_2[L_2] + z$, with a_j nonnegative, $z \in \overline{NE}(X)$ and $Q \cdot z \geq 0$. If neither R_1 nor R_2 is extremal, write $[L_i] = r_i + s_i$, where the ray $\mathbf{R}^+ r_i$ is extremal, $Q \cdot r_i \geq 0$ and $s_i = a_{i1}[L_1] + a_{i2}[L_2] + z_i$, with a_{ij} nonnegative, $a_{ii} < 1$, $z_i \in \overline{NE}(X)$ and $Q \cdot z_i \geq 0$. We get

$$\left(1 - a_{22} - \frac{a_{12}a_{21}}{1 - a_{11}}\right)[L_2] = \frac{a_{21}}{1 - a_{11}}(r_1 + s_1) + (r_2 + s_2)$$

Intersecting with Q, we see that the coefficient of $[L_2]$ is negative. Since $\overline{NE}(X)$ contains no lines, r_2 must vanish, which is a contradiction.

So say R_1 is extremal. If its contraction exists, it is by [Mo2], th. (3.3), the blow-up of a smooth variety along a smooth rational curve, whose normal bundle has to be $\mathcal{O}(-1) \oplus \mathcal{O}(-1)$ (use footnote 9, p. 162).

The curves Γ_i are all algebraically equivalent (they are fibers of the $\mathbf{P}^1$-bundle $S \to S'$), hence have the same class $[\Gamma]$. Let $\alpha = \varepsilon' \circ \varepsilon$. The relative effective cone $\mathrm{NE}(\alpha)$ is generated by the classes $[\Gamma]$, $[L]$, and $[E_i]$. Since the vector space $N_1(X)_\mathbf{R}/\alpha^* N_1(X'')_\mathbf{R}$ has dimension 2, there must be a relation

$$E_i \sim a_i L + b_i \Gamma$$

Using the formulas given in 6.15 for smooth blow-ups, we get

$$\begin{array}{ll} K_X \cdot \Gamma = -1 & K_X \cdot L = K_{X'} \cdot \varepsilon(L) = -2 \\ C \cdot E_i = C' \cdot E_i' = -1 & K_X = \alpha^* K_{X''} + S + 2C \end{array}$$

On the other hand, since $C = \varepsilon^* C'$ and S and L are disjoint, we also have

$$C \cdot \Gamma = S \cdot L = 0$$

hence $S \cdot \Gamma = -1$ and $C \cdot L = -1$. Finally, since S and P_i meets transversally in E_i, we have $S \cdot E_i = 1$. This implies $a_i = -b_i = 1$ and the E_i are all numerically equivalent to $L - \Gamma$. The cone $\mathrm{NE}(\alpha)$ is therefore generated by $[\Gamma]$ and $e = [L - \Gamma]$. Since it is an extremal subcone of $\mathrm{NE}(X)$, the class e spans a K_X-negative (one has $K_X \cdot e = -1$) extremal ray which can be contracted.[8] The corresponding contraction $c : X \to Y$ maps each P_i to a point. Its exceptional locus is the disjoint union $P_1 \sqcup \cdots \sqcup P_r$.

6.20. A flip in dimension 3.

We assume that the characteristic is zero, so that negative extremal rays can be contracted by Theorem 7.39.

We start from the end product of the flip, which is a smooth variety X^+ containing a smooth rational curve Γ^+ with normal bundle $\mathscr{O}(-1) \oplus \mathscr{O}(-2)$, such that the K_{X^+}-positive ray $\mathbf{R}^+[\Gamma^+]$ can be contracted by a morphism $X \to Y$ (take, for example, $X^+ = \mathbf{P}(\mathscr{O}_{\mathbf{P}^1} \oplus \mathscr{O}_{\mathbf{P}^1}(1) \oplus \mathscr{O}_{\mathbf{P}^1}(2))$ and take for Γ^+ the image of the section of the projection $X^+ \to \mathbf{P}^1$ corresponding to the trivial quotient of $\mathscr{O}_{\mathbf{P}^1} \oplus \mathscr{O}_{\mathbf{P}^1}(1) \oplus \mathscr{O}_{\mathbf{P}^1}(2)$; it is contracted by the base-point-free linear system $|\mathscr{O}_{X^+}(1)|$).

Let $X_1^+ \to X^+$ be the blow-up of Γ^+. The exceptional divisor is the ruled surface

$$S_1^+ = \mathbf{P}(N_{\Gamma^+/X^+}) = \mathbf{P}(\mathscr{O}_{\mathbf{P}^1} \oplus \mathscr{O}_{\mathbf{P}^1}(-1))$$

which has a section E_1^+ with self-intersection -1, whose normal bundle in X_1^+ is isomorphic to $\mathscr{O}(-1) \oplus \mathscr{O}(-1)$.[9] Blow-up the curve E_1^+ in X_1^+ to get

---

[8] In characteristic zero, this follows from Theorem 7.39. It can also be seen directly (in all characteristics) by showing that if H is an ample divisor on X'', the divisor $m\alpha^* H - S - C$ is base-point-free for all m big enough and defines the contraction.

[9] Let C be a smooth curve in a smooth threefold Y and let $\varepsilon : X \to Y$ be its blow-up, with exceptional divisor $S = \mathbf{P}(N_{C/Y})$. Let C_0 be a smooth section of $S \to C$. Taking the degrees in the exact sequence

$$0 \to N_{C_0/S} \to N_{C_0/X} \to (N_{S/X})|_{C_0} \to 0$$

a smooth threefold X_0. The exceptional divisor is now the ruled surface $S_0 = \mathbf{P}^1 \times \mathbf{P}^1$, and its normal bundle is of type $(-1, -1)$. Let Γ_0 be a fiber of $S_0 \to E_1^+$. A section is given by the intersection of the strict transform of S_1^+ (which we will still denote by S_1^+) with S_0, which we will also denote by E_1^+.

The notation is summarized in the following diagram.

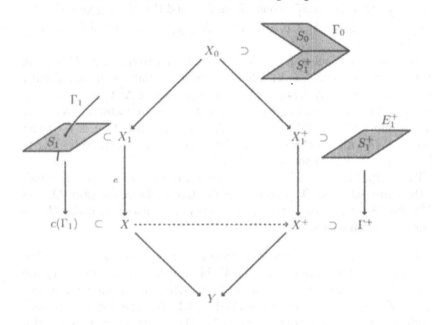

A flip.

The K_{X_0}-*negative ray* $\mathbf{R}^+[E_1^+]$ *is extremal.* Indeed, the relative cone of the morphism $X_0 \to X_1^+ \to X^+ \to Y$, generated by $[E_1^+]$, $[\Gamma_0]$, and the class of the strict transform F_0 of a fiber of $S_1^+ \to \Gamma^+$, is extremal by Proposition 1.14(a). If $\mathbf{R}^+[E_1^+]$ is *not* extremal, one can therefore write $[E_1^+] = a[F_0] + b[\Gamma_0]$ with a and b positive. Intersecting with S_0, we get $-1 = a - b$. Intersecting with (the strict transform of) S_1^+, we get the relation $0 = -a + b$, which is absurd.

---

yields $-K_X \cdot C_0 + \deg(K_{C_0}) = C_0^2 + S \cdot C_0$ (where C_0^2 is computed in the surface S). The adjunction formula $K_X = \varepsilon^* K_Y + S$ yields on the other hand $K_X \cdot C_0 = K_Y \cdot C_0 + S \cdot C_0$. All this implies

$$C_0^2 + 2S \cdot C_0 = \deg(N_{C/Y})$$

If C is rational, we have an exact sequence

$$0 \to \mathscr{O}_{C_0}(C_0^2) \to N_{C_0/X} \to \mathscr{O}_{C_0}(\tfrac{1}{2}(\deg(N_{C/Y}) - C_0^2)) \to 0$$

which is split if $\deg(N_{C/Y}) \geq 3C_0^2 - 2$.

As explained in 6.18, its contraction is the blow-up of a smooth threefold X_1 along a smooth rational curve Γ_1 with normal bundle $\mathcal{O}(-1) \oplus \mathcal{O}(-1)$, so that $K_{X_1} \cdot \Gamma_1 = 0$. The exceptional curve E_1^+ of S_1^+ gets blown down so S_1^+ maps onto a projective plane S_1.

To compute the normal bundle to S_1 in X_1, we restrict to a line F_1 in S_1 which does not meet Γ_1. This restriction is the same as the restriction of $N_{S_1^+/X_0}$ to a line in S_1^+ disjoint from E_1^+, and this has degree $(-1 - 2 - 1)/2 = -2$ by footnote 9, p. 162, hence $N_{S_1/X_1} \simeq \mathcal{O}(-2)$ and $(K_{X_1})|_{S_1} \simeq \mathcal{O}(-1)$.

We have in particular, $K_{X_1} \cdot F_1 = -1$, and the extremal ray $\mathbf{R}^+[F_1]$ can be contracted by $c : X_1 \to X$. A local study shows that locally analytically at $c(S_1)$, the variety X is isomorphic to the quotient of $\mathbf{A}^3$ by the involution $x \mapsto -x$. The corresponding complete local ring is *not* factorial: its divisor class group ([H1], p. 131) has order 2. It follows that $2K_X$ is a Cartier divisor. Write $K_{X_1} = c^* K_X + a[S_1]$, for some rational a. By restricting to S_1, we get $a = 1/2$, hence $K_X \cdot c(\Gamma_1) = -1/2$.

The morphism $X \to Y$ is the contraction of the ray $\mathbf{R}^+[c(\Gamma_1)]$, which is therefore extremal. The corresponding flip is the composition $X \dashrightarrow X^+$: the "K_X-negative" rational curve $c(\Gamma_1)$ is replaced with the "K_{X^+}-positive" rational curve Γ^+.

Divisorial contractions on smooth threefolds have been classified by Mori in [Mo2], th. (3.3). Apart from smooth blow-ups, he shows that they are either of the type $c : X_1 \to X$ above (contraction of a plane with normal bundle $\mathcal{O}(-2)$), or of the type studied in 6.18 (contraction of a smooth quadric with normal bundle of type $(-1, -1)$), or the contraction of a singular quadric with normal bundle $\mathcal{O}(-1)$.

Small contractions on smooth fourfolds are described in [Ka5]: the exceptional locus is the disjoint union of copies of $\mathbf{P}^2$ (see 6.19) whose normal bundle is $\mathcal{O}_{\mathbf{P}^2}(-1) \oplus \mathcal{O}_{\mathbf{P}^2}(-1)$, and the flip exists. Note that if R is the corresponding extremal ray, it follows already from (6.6) that R has length 1 and that each connected component of its locus is contracted to a point by c_R.

Extremal contractions on smooth fourfolds are described in [AW1] and [AW2].

6.7 Exercises

1. Let X be a smooth projective surface and let R be a K_X-negative extremal ray in $N_1(X)_{\mathbf{R}}$. By the cone theorem 6.1, R is generated by the class of a rational curve Γ on X such that

$$K_X \cdot \Gamma \in \{-1, -2, -3\}$$

Using Lemma 6.2, show that the contraction c_R of R exists and that one of the following occurs:

(a) either $\Gamma^2 > 0$ and the image of c_R is a point;

(b) or $\Gamma^2 = 0$, the surface X is ruled over a smooth curve C, and c_R is the corresponding map $X \to C$;

(c) or $\Gamma^2 < 0$, the curve Γ is exceptional, and c_R is the blow-down of that curve.

2. Let X be a smooth projective variety and let M be a nef divisor on X. Show that $K_X + tM$ is nef for all $t \geq \dim(X) + 1$ (see Exercise 6.7.5 for a better result).

3. Let X be a smooth projective Fano variety of positive dimension n, let $f : \mathbf{P}^1 \to X$ be a rational curve of $(-K_X)$-degree $\leq n+1$ (which exists by Theorem 3.4), let M be a component of $\mathrm{Mor}(\mathbf{P}^1, X; 0 \mapsto f(0))$ containing $[f]$, and let

$$\mathrm{ev} : \mathbf{P}^1 \times M \to X$$

be the evaluation map. Assume that the $(-K_X)$-degree of any rational curve on X is $\geq (n + 3)/2$.

(a) Show that $Y = \mathrm{ev}(\mathbf{P}^1 \times M)$ is closed in X and that its dimension is at least $(n + 1)/2$.

(b) Show that any curve in Y is algebraically equivalent in Y to some rational multiple of $f(\mathbf{P}^1)$.

(c) Conclude that $N_1(X)_{\mathbf{R}}$ has dimension 1.

4. **Nonisomorphic minimal models in dimension 3.** Let S be a Del Pezzo surface, i.e., a smooth Fano surface. Set

$$P = \mathbf{P}(\mathscr{O}_S \oplus \mathscr{O}_S(-K_S)) \xrightarrow{\pi} S$$

and let S_0 be the image of the section of π that corresponds to the trivial quotient of $\mathscr{O}_S \oplus \mathscr{O}_S(-K_S)$, so that the restriction of $\mathscr{O}_P(1)$ to S_0 is trivial.

(a) What is the normal bundle to S_0 in P?

(b) By considering a cyclic cover of P branched along a suitable section of $\mathscr{O}_P(m)$, for m large, construct a smooth projective threefold of general type X with K_X nef that contains S as a hypersurface with normal bundle K_S.

(c) Assume from now on that S contains an exceptional curve C (i.e., a smooth rational curve with self-intersection -1). What is the normal bundle of C in X?

(d) Let $\tilde{X} \to X$ be the blow-up of C. Describe the exceptional divisor E.

(e) Let C_0 be the image of a section $E \to C$. Show that the ray $\mathbf{R}^+[C_0]$ is extremal and $K_{\tilde{X}}$-negative.

(f) Assume moreover that the characteristic is zero. The ray $\mathbf{R}^+[C_0]$ can be contracted (according to Theorem 7.39) by a morphism $\tilde{X} \to X^+$. Show that X^+ is smooth, that K_{X^+} is nef, and that X^+ is not isomorphic to X. The induced rational map $X \dashrightarrow X^+$ is called a *flop*.

5. **A rationality theorem.** Let X be a smooth projective variety whose canonical divisor is not nef and let M be a nef divisor on X. Set

$$r = \sup\{t \in \mathbf{R} \mid M + tK_X \text{ nef}\}$$

(a) Let $(\Gamma_i)_{i \in I}$ be the (nonempty and countable) set of rational curves on X that appears in the cone theorem 6.1. Show

$$r = \inf_{i \in I} \frac{M \cdot \Gamma_i}{-K_X \cdot \Gamma_i}$$

(b) Deduce that one can write

$$r = \frac{u}{v}$$

with u and v relatively prime integers and $0 < v \le \dim(X) + 1$, and that there exists a K_X-negative extremal ray R of $\overline{NE}(X)$ such that

$$(M + rK_X) \cdot R = 0$$

6. **The effective cone of an abelian variety (T. Bauer).** Let X be an abelian variety.

(a) Assume that X is simple (i.e., its only abelian subvarieties are $\{0\}$ and X) and that $\overline{NE}(X)$ is a finitely generated cone. Show that the Picard number of X is 1. (*Hint:* Proceed by contradiction and, letting H and H' be nonproportional ample divisors on X, prove that $r = \inf\{t \in \mathbf{R} \mid tH - H' \text{ nef }\}$ is irrational.)

(b) Assume X has Picard number 1. Show that $\overline{NE}(X \times X)$ is *not* a finitely generated cone.

(c) Show that $\overline{NE}(X)$ is a finitely generated cone if and only if X is isogeneous to a product of mutually nonisogeneous abelian varieties with Picard number 1.

7
Cohomological Methods

We begin this chapter with a discussion of various models that can be attached to a projective variety and explore in much more detail the leads developed in Chapter 1. One idea was to attach to a projective variety X a preferred member $X_{\min}$ of its birational equivalence class with good properties, such as $K_{X_{\min}}$ nef. It is then minimal in the sense that any birational map from $X_{\min}$ to a smooth projective variety is an isomorphism (Proposition 1.45). This is possible only if X is not uniruled and the (explicit) construction of $X_{\min}$ is one of the purposes of Mori's minimal model program.

There is on the other hand an object canonically attached to X, to wit its *canonical algebra*

$$R(X, K_X) = \bigoplus_{m \geq 0} H^0(X, mK_X)$$

This is a birational invariant of X. When X is a smooth projective surface, the algebra $R(X, K_X)$ is known to be finitely generated and

$$X_{\mathrm{can}} = \mathrm{Proj}(R(X, K_X))$$

is called the *canonical model* of X (although it may not be birational to X, all pluricanonical morphisms factor through it). It is one of the main conjectures of higher-dimensional geometry that the algebra $R(X, K_X)$ should always be finitely generated. We prove in Corollary 7.33 that this is the case when K_X is nef and big (i.e., for a minimal model of a variety of general type).

To sum up, we expect the following picture when X is not uniruled:

$$X \dashrightarrow X_{\min} \longrightarrow X_{\text{can}}$$

where the first (conjectural) birational map is furnished by the minimal model program and the second map contracts all curves on which the nef divisor $K_{X_{\min}}$ has degree 0.

The usual method to prove that the algebra $R(X, K_X)$ is finitely generated is based on a result of Zariski (Proposition 7.6), which says that this is the case if some multiple of K_X is base-point-free. This property requires in turn that K_X be nef, so again we see here the importance of the construction of a minimal model.

So back to Mori's program. We know from the last chapter that we need to allow some kind of singularities. We define in Section 7.2 two classes of singular varieties: varieties with canonical singularities and varieties with terminal singularities. The first class arises naturally as the type of singularities which appear on canonical models, whereas the second (smaller) class seems to be the smallest class which is stable under the birational operations of the minimal model program described in Chapter 6 (divisorial contractions and flips). It should therefore be the type of singularities that one finds on minimal models of nonuniruled varieties (in dimension 2, canonical singularities are rational double points and terminal singularities are smooth). We show in 7.18 that birationally equivalent minimal models with terminal singularities are isomorphic in codimension 1 (but they may be nonisomorphic).

Allowing singular varieties in Mori's program means that we need to have theorems that apply not only to smooth varieties (as in Chapter 6), but also to varieties with say canonical singularities. The methods we used in the smooth case, all more or less based on bend-and-break techniques, are completely powerless in the singular case, and we need an entirely new approach. It is based on seemingly innocuous generalizations of the celebrated Kodaira vanishing theorem, due for the most part to Kawamata and Viehweg (Theorems 7.21 and 7.26). We state these theorems without proof in Section 7.3, and in Section 7.4 using the formalism of logarithmic geometry. The drawback is that this approach works only in characteristic zero, an assumption that is therefore necessary for most results in this chapter. It seemed impossible not to mention in this connection the beautiful theory of multiplier ideals and the Nadel vanishing theorem. We only go through the basics of the theory (which is not used anywhere else in the book) in Section 7.5, and refer to the forthcoming book by R. Lazarsfeld ([L]) for more details.

Our aim is to prove the cone theorem for singular varieties. It comes as the last in a series of three difficult theorems, each of independent interest:

- Shokurov's nonvanishing theorem 7.30, which gives conditions on Cartier divisors D and A on a projective variety X under which the linear system $|mD + A|$ is nonempty for all integers m sufficiently large;

- the base-point-free theorem 7.32, which gives conditions on a nef Cartier divisor D on a projective variety X under which the linear system $|mD|$ is base-point-free for all integers m sufficiently large;

- the rationality theorem 7.34, which states that if H is a nef divisor on a projective variety X with canonical singularities, the number $\sup\{t \in \mathbf{R} \mid H + tK_X \text{ nef}\}$ is infinite or *rational*.

The cone theorem is then a formal consequence of the rationality theorem, which allows the construction of nef but nonample divisors. It states that for a projective variety X with canonical singularities, the set $\mathscr{R}$ of all K_X-negative extremal rays of $\overline{\mathrm{NE}}(X)$ is countable and

$$\overline{\mathrm{NE}}(X) = \overline{\mathrm{NE}}(X)_{K_X \geq 0} + \sum_{R \in \mathscr{R}} R$$

The reader will notice that rational curves, in sharp contrast with the smooth case where they are an essential ingredient of the proof via bend-and-break arguments, are not part of the picture here. It is however still true that each ray in $\mathscr{R}$ is generated by the class of a rational curve, but this comes as an independent result (Theorem 7.46) due to Kawamata and obtained via a clever use of bend-and-break.

An essential by-product of the proof of the cone theorem and the base-point-free theorem that comes for free is that K_X-negative extremal sub-cones can be contracted (Theorem 7.39), a result that was out of the reach of the methods of Chapter 6. We discuss the resulting morphisms and define flips in Section 7.10. We prove that the class of projective varieties with canonical (resp. terminal) singularities is stable under divisorial contractions and flips. One of the main remaining unsolved problems of the theory is the existence of flips and their termination (i.e., there cannot be an infinite chain of flips). This is known in dimension 3.

I have chosen to state and prove these theorems within the framework of *logarithmic geometry*, where instead of a single projective variety X, one considers a pair (X, Δ), where Δ is a "boundary" effective $\mathbf{Q}$-divisor (the original idea was that many properties of the open variety $X - \Delta$ depend on the $\mathbf{Q}$-divisor $K_X + \Delta$, but this point of view is not apparent here). This minor change makes the statements look much more natural and brings spectacular improvements to the theory, even in the "standard" case (see, for example, Exercises 7.13.8 and 7.13.9), at minimal cost because the proofs of the relevant results are virtually identical. This extension is also important because it provides the opportunity to introduce the concept of *singularity of a pair*. I stick to the minimal amount of machinery here, and only define *log terminal* pairs, which satisfy the exact condition that makes the proof of the vanishing theorem (and of its consequences) work. The pair consisting of a variety with canonical singularities and the zero divisor is a log terminal pair (in my opinion, the confusing terminology is to a large part responsible for the undeserved opacity and confidentiality of the theory).

In the last section, we briefly discuss another important extension of the cone theorem and the minimal model program to the relative case, where we have a projective morphism $\pi : X \to S$ and all the constructions are over S. We still impose conditions on the singularities of X and study properties of its canonical divisor, but are interested in the structure of the relative closed cone of curves $\overline{\mathrm{NE}}(\pi)$, as defined in 1.12, and S-contractions.

Both extensions can be done simultaneously, at little extra cost. This is the point of view taken in the reference texts [KaMM], [KM], and [M], but the beginner in the field will probably be disheartened by the technicalities, which have the unfortunate effect of hiding the really important ideas. It seems preferable to understand fully the simplest case before moving on to higher spheres.

Of course, there are many important points of the theory that are not covered here, and in particular the three-dimensional case, where most conjectures are now theorems (see [K8] for a complete treatment of this case). This is mainly due to lack of space and competence. I have certainly not tried to be exhaustive (who could?) but rather to give the reader a taste for this remarkable theory and for further reading.

All schemes and varieties are defined over an algebraically closed field $\mathbf{k}$, which will be of characteristic zero as of Section 7.3.

7.1   Canonical Models

Given a smooth projective variety X, we want to define a "model" of X that depends only on the birational equivalence class of X. One of the objects intrinsically attached to X is the $\mathbf{k}$-algebra

$$R(X, K_X) = \bigoplus_{m \geq 0} H^0(X, mK_X)$$

which is called the *canonical algebra* of X.

7.1. It is indeed a birational invariant: by 1.39, a birational morphism $u : X \dashrightarrow X'$ between smooth projective varieties is defined on an open set U whose complement has codimension at least 2. The dominant morphism $u|_U : U \to X'$ induces an injection $u^* : H^0(X', mK_{X'}) \to H^0(U, mK_U)$, and the restriction $H^0(X, mK_X) \to H^0(U, mK_U)$ is bijective ([H1], III, ex. 3.5). We may therefore define an injective morphism

$$u^* : H^0(X', mK_{X'}) \to H^0(X, mK_X)$$

whose inverse is $(u^{-1})^*$.

7.2. Key problem: *is the* k-*algebra* $R(X, K_X)$ *finitely generated?*

7.3. This is a fundamental question: assuming the answer is yes, the projective variety

$$X_{\text{can}} = \text{Proj}(R(X, K_X))$$

depends only on the birational equivalence class of X. It is called the *canonical model* of X (be careful that this variety may not be birationally equivalent to X: after all, it is empty if X has no nonzero pluricanonical forms!). It may be concretely realized as follows: for any positive integer r, set

$$R_r = H^0(X, rK_X), \qquad R^{(r)} = \bigoplus_{m \geq 0} R_{mr}$$

There exists an integer r such that $R^{(r)}$ is generated by R_r, which means that the morphism

$$\text{Sym}\, R_r \to R^{(r)}$$

is surjective. The variety X_{can} is then isomorphic to the subvariety of $\mathbf{P}(R_r)$ defined[1] by the kernel of this morphism, which is also the image of the rational map

$$\varphi_{rK_X} : X \dashrightarrow \mathbf{P}(R_r)$$

associated with the complete linear system $|rK_X|$ ([H1], II, th. 7.1). In particular, there is a canonical rational map $\varphi_{\text{can}} : X \dashrightarrow X_{\text{can}}$ such that

- any *pluricanonical map* φ_{mK_X} factors as

$$\varphi_{mK_X} : X \xrightarrow{\varphi_{\text{can}}} X_{\text{can}} \dashrightarrow \mathbf{P}(R_m)$$

- the dimension of X_{can} is $\max_{m>0} \dim(\varphi_{mK_X}(X))$.

The first item can be seen as follows: for any positive m, the rational map φ_{mrK_X} is the composition of φ_{can} with the m-uple Segre embedding ([H1], II, ex. 5.13). Similarly, the composition of φ_{mK_X} with the r-uple Segre embedding is the composition of φ_{mrK_X} with the rational map $\mathbf{P}(R_{mr}) \dashrightarrow \mathbf{P}(\text{Sym}^r R_m)$ coming from the morphism $\text{Sym}^r R_m \to R_{mr}$.

Example 7.4 A smooth projective surface of general type X has a unique minimal model X_{min} obtained by contracting on X all smooth rational curves with self-intersection -1. The surface X_{min} is smooth and the divisor $K_{X_{\text{nef}}}$ is nef, but has degree 0 on smooth rational curves with self-intersection -2. The canonical model X_{can} is the surface obtained from X_{min} by contracting these curves. It has at worst rational double points and is isomorphic to the image of φ_{5K_X}.

---

[1] As usual, we follow Grothendieck's notation: the projectivization $\mathbf{P}(V)$ of a vector space V is the space of *hyperplanes* in V.

7.5. Going back to the key problem 7.2, note that for any positive integer r, *the* **k**-*algebra* R *is finitely generated if and only if the* **k**-*algebra* $R^{(r)}$ *is finitely generated.*

Indeed, assume that $R^{(r)}$ is finitely generated. Note that the **k**-algebra R embeds into $k(X)[t]$ by sending an element s of R_m to st^m. It follows that its quotient field is a subfield of $k(X)(t)$, hence has transcendence degree at most $\dim(X) + 1$ over **k**. Now R is an integral extension of $R^{(r)}$, hence is finitely generated by finiteness of the integral closure (see [H1], I, th. 3.9A). Conversely, if R is finitely generated, some $R^{(s)}$ is a quotient of the polynomial algebra $\operatorname{Sym} R_s$, and $R^{(rs)}$ is a quotient of $\operatorname{Sym} R_{rs}$, hence is finitely generated. By what we just proved, so is $R^{(r)}$.

The current methods for attacking the key problem 7.2 are based on the following result of Zariski.

Proposition 7.6 *Let* X *be a normal projective variety and let* D *be a Cartier divisor on* X *such that the linear system* $|D|$ *has no base-points.*

(a) *The* **k**-*algebra* $R(X, D) = \bigoplus_{m \geq 0} H^0(X, mD)$ *is finitely generated.*

(b) *For each* r *large enough, the morphism* φ_{rD} *has connected fibers and its image is normal.*

PROOF. Let
$$\varphi_D : X \xrightarrow{\varphi} X' \xrightarrow{p} \mathbf{P}(H^0(X, D))$$
be the Stein factorization of φ_D, where X' is normal, p is finite, and $\varphi_* \mathcal{O}_X \simeq \mathcal{O}_{X'}$ (so that φ has connected fibers). If H is a hyperplane in $\mathbf{P}(H^0(X, D))$, one has, for any positive integer r,

$$
\begin{aligned}
H^0(X, rD) &= H^0(X, \varphi^*(rp^*H)) \\
&\simeq H^0(X', \varphi_* \mathcal{O}_X \otimes \mathcal{O}_{X'}(rp^*H)) \\
&\simeq H^0(X', rp^*H)
\end{aligned}
$$

In other words, the morphism φ_{rD} factors as $\varphi_{rp^*H} \circ \varphi$. Since rp^*H is very ample for r large enough, φ_{rp^*H} is then an embedding. This proves (b).

Identify X' with a subvariety of $\mathbf{P}(H^0(X', rp^*H))$ by means of the closed embedding φ_{rp^*H}, and let H' be a hyperplane in that projective space. The **k**-algebra $R(X, D)^{(r)}$ is isomorphic to

$$\bigoplus_{m \geq 0} H^0(X', mH') = R(X', H')$$

By 1.19, the higher cohomology of $I_{X'}(s)$ vanishes for all s large enough. The restrictions

$$\operatorname{Sym}^{ms} H^0(X', H') \to H^0(X', msH')$$

are then surjective for all m, hence the k-algebra $R(X', H')^{(s)}$ is finitely generated, and so are the k-algebras $R(X', H')$ and $R(X, D)$ by 7.5.   □

7.7. The strategy for solving the key problem 7.2 is now the following:

- find a birational model $X_{\min}$ of X, called a *minimal model*, with $K_{X_{\min}}$ *nef*. We also require $X_{\min}$ to have mild singularities called *terminal* (see Definition 7.13);

- prove that some multiple of $K_{X_{\min}}$ has no base-points.

The first step is the object of Mori's minimal model program. The second step is usually called the *abundance conjecture*.

Note that $R(X, K_X)$ might well be 0. This is the case in characteristic zero if X is uniruled by Corollary 4.12 (and it is conjectured to be the only case where this happens; see 4.13). Also, if X is uniruled, it cannot have a minimal model, since it would contradict Example 4.7(1) and Corollary 4.11 (the possible singularities of $X_{\min}$ are not a problem because they occur in codimension at least 3; see Remark 7.17(2)). So the above strategy is conjectured to work only for nonuniruled varieties (and it does in dimension at most 3).

A given variety X (of dimension at least 3) may have nonisomorphic minimal models[2] (see Exercise 6.7.4) so $X_{\min}$ is not in general a birational invariant of X. On the other hand, X_{can} can be a birational model of X only if it has the same dimension, i.e., if X is *of general type* (see footnote 11, p. 75), and Corollary 7.33 shows that the second step above works. By Proposition 7.6(b), the corresponding morphism $X_{\min} \to X_{\text{can}}$ is birational onto its image, and X_{can} is a birational model of X.

7.8. What is known? The answer to the key problem is expected to be affirmative in all cases. This is known for minimal models of projective varieties of general type of any dimension by Theorem 7.32 (i.e., for projective varieties with canonical singularities whose canonical divisor is nef and big). It is also known in dimension at most 3, because the minimal model program (first step in 7.7) works and the abundance conjecture holds (see [K8]).

---

[2]The examples of [Fr], which are essentially obtained by taking double covers of the various varieties appearing in 6.20, are not minimal in our sense (the canonical divisor is not nef).

In dimension 3, there is a precise description of how various minimal models are related: any birational map between two such models is a composition of elementary operations called *flops* (see [K6], th. 4.9). A flop removes a curve on which the canonical divisor has degree 0 and replaces it with another curve with the same property. However, there is a Cartier divisor that is negative on the first curve and positive on the second (compare with flips, defined in 6.13 and Definition 7.43).

7.2 Singularities

We now turn to the study of singularities of canonical models (aptly named canonical singularities). Since these singularities are characterized in terms of a desingularization, we will assume that the characteristic of the base field is zero.

For any projective variety X that is nonsingular in codimension 1, we define a Weil divisor class K_X as the class of the closure of a canonical divisor on the smooth locus of X.

Proposition 7.9 *Let X be a smooth projective variety of general type whose canonical algebra $R(X, K_X)$ is finitely generated.*[3] *Set*

$$X_{\mathrm{can}} = \mathrm{Proj}(R(X, K_X))$$

(a) *The variety X_{can} is normal.*

(b) *Some multiple of $K_{X_{\mathrm{can}}}$ is an ample Cartier divisor.*

(c) *For any desingularization $\pi : Y \to X_{\mathrm{can}}$, there exists an effective $\mathbb{Q}$-divisor F such that*

$$K_Y \sim \pi^* K_{X_{\mathrm{can}}} + F$$

Let us begin with an auxiliary result of independent interest. It is stated in a setting slightly more general than necessary here, because it is no more complicated and will be useful later on.

Let X be a *normal* projective variety and let D be a Cartier divisor on X. Assume that the algebra $R(X, D)$ is finitely generated and set $X_D = \mathrm{Proj}(R(X, D))$. The discussion of 7.3 applies: there is a positive integer r such that X_D is isomorphic to the image of $\varphi = \varphi_{rD} : X \dashrightarrow \mathbf{P}(H^0(X, rD))$ and any map φ_{mD} factors through φ.

Lemma 7.10 *Let X be a normal projective variety and let D be a Cartier divisor on X. Assume that the algebra $R(X, D)$ is finitely generated.*

(a) *The variety $X_D = \mathrm{Proj}(R(X, D))$ is normal.*

(b) *If X_D has same dimension as X, there exist an open subset X^0 of X, and an open subset X_D^0 of X_D whose complement has codimension at least 2 in X_D, between which $\varphi|_{X^0}$ induces an isomorphism. Some multiple of $(\varphi|_{X^0}^{-1})^* D$ is the restriction of an ample Cartier divisor of X_D to X_D^0.*

---

[3]By Theorem 7.32 and Proposition 7.6, this condition holds if K_X is nef.

PROOF. Let r be the integer defined above and let X' be the normalization of the graph of φ, so that there is a birational morphism $\varepsilon : X' \to X$ such that

$$\varphi' = \varphi \circ \varepsilon : X' \twoheadrightarrow X_D \subset \mathbf{P}(H^0(X, rD))$$

is a morphism. If $|M|$ is the associated base-point-free linear system on X', we have

$$|r\varepsilon^* D| = |M| + F$$

where F is an effective Cartier divisor on X' which is the base-locus of $|r\varepsilon^* D|$. The map ε is an isomorphism outside the base-locus of $|rD|$, hence the exceptional locus of ε is contained in the support of F.

For any positive integer m, we have a commutative diagram

$$
\begin{array}{ccc}
\operatorname{Sym}^m H^0(X, rD) & \twoheadrightarrow & H^0(X, mrD) \\
\downarrow{\scriptstyle \varepsilon^*} & & \downarrow{\scriptstyle \varepsilon^*} \\
\operatorname{Sym}^m H^0(X', r\varepsilon^* D) & \longrightarrow & H^0(X', mr\varepsilon^* D) \\
\| & & \uparrow{\scriptstyle \alpha} \\
\operatorname{Sym}^m H^0(X', M) & \longrightarrow & H^0(X', mM)
\end{array}
$$

where, by the following lemma, the maps ε^* are bijective. It follows that the injective morphism α is bijective. In particular, φ_{mM} factors as $\varphi_{mrD} \circ \varepsilon$, and it follows from Proposition 7.6(b) that X_D is normal.

Lemma 7.11 *Let X and Y be varieties, with X normal, and let $\pi : Y \to X$ be a proper birational morphism. Let D be a Cartier divisor on X and let F an effective Cartier divisor on Y whose support is contained in $\operatorname{Exc}(\pi)$. We have*

$$H^0(X, D) \simeq H^0(Y, \pi^* D + F)$$

We will prove this lemma after the end the proof of Proposition 7.9 and proceed presently with the proof of item (b) of Lemma 7.10.

We now assume that φ is birational and prove (b). The low-degree terms of the Leray spectral sequence for φ' fit into an exact sequence ([Br], IV.6)

$$0 \to H^1(X_D, \varphi'_* \mathcal{O}_{X'} \otimes \mathcal{O}(m)) \to H^1(X', mM) \to H^0(X_D, R^1 \varphi'_* \mathcal{O}_{X'} \otimes \mathcal{O}(m))$$

By [H1], III, cor. 11.2, the support of $R^1 \varphi'_* \mathcal{O}_{X'}$ is contained in the set of points of X_D whose fiber has dimension at least 1; hence it has codimension at least 2 in X_D since φ' is birational. On the other hand, $H^1(X_D, \varphi'_* \mathcal{O}_{X'} \otimes \mathcal{O}(m))$ vanishes for all m large enough (1.19). Letting n be the dimension of X (and of X_D), Proposition 1.31 implies

$$h^1(X', mM) = O(m^{n-2})$$

because M is nef and big. By Lemma 7.11, F is the base-locus of the linear system $|mM + F|$. In the exact sequence

$$H^0(X', mM + F) \xrightarrow{\rho} H^0(F, mM + F) \to H^1(X', mM)$$

the map ρ therefore vanishes, hence $h^0(F, mM + F) = O(m^{n-2})$. Write $F|_F \equiv D'' - D'$, where D' and D'' are effective Cartier divisors on F. The exact sequence

$$0 \to H^0(F, mM - D') \to H^0(F, mM) \to H^0(D', mM)$$

implies

$$
\begin{aligned}
h^0(\varphi'(F), \mathcal{O}(m)) &\leq h^0(F, mM) \\
&\leq h^0(F, mM - D') + h^0(D', mM) \\
&\leq h^0(F, mM + F) + h^0(D', mM) \\
&= O(m^{n-2})
\end{aligned}
$$

It follows that the dimension of $\varphi'(F)$ is at most $n - 2$.

Consider the closed subset of X_D of codimension at least 2 which is the union of the singular locus of X_D and $\varphi'(F \cup \mathrm{Exc}(\varphi'))$. Its complement X_D^0 is smooth, φ' induces an isomorphism between its inverse image X'^0 and X_D^0, hence φ induces an isomorphism between the (smooth) open subset $X^0 = \varepsilon(X'^0)$ of X and X_D^0. Moreover, if H is a hyperplane in $\mathbf{P}(H^0(X, rD))$, we have

$$
\begin{aligned}
(\varphi|_{X^0})^* H &\equiv (\varepsilon^{-1})^* (\varphi'|_{X'^0})^* H \\
&\equiv (\varepsilon^{-1})^* (M|_{X'^0}) \\
&\equiv (\varepsilon^{-1})^* ((M + F)|_{X'^0}) \\
&\equiv (rD)|_{X^0}
\end{aligned}
$$

This proves (b). $\square$

PROOF OF THE PROPOSITION. Lemma 7.10 implies items (a) and (b) of the proposition. Let us prove (c), keeping the same notation as above. Let $\pi : Y \to X_{\mathrm{can}}$ be a desingularization. Upon replacing X' by the normalization of the graph of the rational map $X' \dashrightarrow Y$, we may assume that there is a factorization

$$\varphi' : X' \xrightarrow{\psi} Y \xrightarrow{\pi} X_{\mathrm{can}} \subset \mathbf{P}(H^0(X', M))$$

On the (smooth) complement X_1' of $\mathrm{Exc}(\psi)$ in X', we have

$$
\begin{aligned}
\psi^*(\pi^*(rK_{X_{\mathrm{can}}})) &\equiv M \\
&\equiv \varepsilon^*(rK_X) - F \\
&\equiv rK_{X_1'} - r\,\mathrm{Ram}(\varepsilon|_{X_1'}) - F \\
&\equiv r\psi^* K_Y - r\,\mathrm{Ram}(\varepsilon|_{X_1'}) - F
\end{aligned}
$$

Transplanting this equivalence by the isomorphism $\psi|_{X_1'}$ to an open subset of Y whose complement has codimension at least 2, we get (c).   □

We still need to prove Lemma 7.11.

PROOF OF THE LEMMA. Recall that X and Y are varieties, X is normal, $\pi : Y \to X$ is a proper birational morphism, D is a Cartier divisor on X, and F is an effective Cartier divisor on Y whose support is contained in the exceptional locus E of π. We have the following chain of inclusions

$$H^0(X, D) \subset H^0(Y, \pi^*D) \subset H^0(Y, \pi^*D + F) \subset H^0(Y - E, \pi^*D + F)$$

and of isomorphisms

$$H^0(Y - E, \pi^*D + F) \simeq H^0(Y - E, \pi^*D) \simeq H^0(X - \pi(E), D) \simeq H^0(X, D)$$

the last one holding because X is normal and $\pi(E)$ has codimension at least 2 in X (1.40 and [H1], III, ex. 3.5). All these spaces are therefore isomorphic, hence the lemma.   □

7.12. A few comments about Lemma 7.11 will be helpful later on. Its conclusion can be rephrased as

$$\pi_*(\mathcal{O}_Y(F)) \simeq \mathcal{O}_X \tag{7.1}$$

Assume Y is smooth. If F is *any* exceptional divisor on Y, we may write $F = F_1 - F_2$, where F_1 and F_2 are effective exceptional divisors with no common component. The sheaf $\pi_*(\mathcal{O}_Y(F))$ is contained in $\pi_*(\mathcal{O}_Y(F_1))$, which is isomorphic to $\mathcal{O}_X$ by (7.1). It is therefore a sheaf of ideals that defines a subscheme of X supported on $\pi(F_2)$. It follows that

$$\pi_*(\mathcal{O}_Y(F)) \simeq \mathcal{O}_X$$

if and only if F is effective.
Still assuming Y smooth, *if D and D' are Cartier divisors on X and F and F' are exceptional divisors on Y (not necessarily effective) such that*

$$\pi^*D + F \equiv \pi^*D' + F'$$

we have $F = F'$ and $D \equiv D'$. Indeed, writing $F = F_1 - F_2$ and $F' = F_1' - F_2'$ as above, we have

$$\pi^*(D - D') + F_1 + F_2' \equiv F_2 + F_1'$$

Taking π_* and using the projection formula and (7.1), we get $D \equiv D'$, hence $F_1 + F_2' = F_2 + F_1'$ by Lemma 7.11. It follows $F_1 = F_1'$ and $F_2 = F_2'$.
Proposition 7.9 motivates the following definition.

Definition 7.13 *A variety X has* canonical singularities *if*

(a) *X is normal;*

(b) *there exists a positive integer j such that jK_X is a Cartier divisor;*

(c) *for some desingularization $\pi : Y \to X$, any j-canonical form on X_{reg} extends to a j-canonical form on Y.*

Of course, a smooth variety has canonical singularities.

7.14. The index of a variety. When it exists, the smallest positive integer j that satisfies (b) is called the *index*[4] of X, written j_X.

Going back to Definition 7.13, if we write

$$jK_Y \equiv \pi^*(jK_X) + F \tag{7.2}$$

where F is a divisor supported in $\text{Exc}(\pi)$ (although this is an equality of divisor classes, 7.12 implies that the divisor F is uniquely defined), the singularities are canonical if and only if F is effective.

This condition is also equivalent to

$$\pi_*(\mathcal{O}_Y(jK_Y)) \simeq \mathcal{O}_X(jK_X) \tag{7.3}$$

Indeed, by the projection formula,

$$\pi_*(\mathcal{O}_Y(jK_Y)) \simeq \mathcal{O}_X(jK_X) \otimes \pi_*(\mathcal{O}_Y(F))$$

and by 7.12, the sheaf $\pi_*(\mathcal{O}_Y(F))$ is trivial if and only if F is effective.

Moreover, if property (c) holds for one desingularization, it holds for all desingularizations. Indeed, let $\pi' : Y' \to X$ be another desingularization, let Z be a desingularization of the graph of the birational map $Y \dashrightarrow Y'$, and let $\tau : Z \to Y$ and $\tau' : Z \to Y'$ be the projections. Writing

$$jK_{Y'} \equiv \pi'^*(jK_X) + F' \quad \text{and} \quad K_Z \equiv \tau^* K_Y + G \equiv \tau'^* K_{Y'} + G'$$

as in (7.2) with G and G' effective (because Y and Y' are smooth), we get

$$jK_Z - (\pi \circ \tau)^*(jK_X) \equiv \tau^*(F) + jG \equiv \tau'^*(F') + jG'$$

Since all these divisors are $(\pi \circ \tau)$-exceptional, it follows from 7.12 that $\tau^* F + G$ and $\tau'^* F' + jG'$ are equal. In particular, if F is effective, $\tau'^* F' + jG'$ is effective, and so is its image F' by τ' (alternatively, one could use (7.3)).

Going back to the setting of Definition 7.13, we say that X has *terminal singularities* if in item (c), any j-canonical form on X_{reg} extends to an

---

[4]It is not to be confused with the Fano index ι_X of a smooth Fano variety X, as defined in Section 5.7.

j-canonical form on Y *with poles at every exceptional divisor of* Y. If we write

$$jK_Y \equiv \pi^*(jK_X) + F$$

as in (7.2), we want $F - \mathrm{Exc}(\pi)^{(1)}$ to be effective, where $\mathrm{Exc}(\pi)^{(1)}$ denote the union of all (reduced) π-exceptional hypersurfaces in Y. Again, this property does not depend on the choice of π (the proof is analogous), and a smooth variety has terminal singularities. In the spirit of (7.3), terminal singularities are characterized by the relation

$$\pi_*(\mathscr{O}_Y(jK_Y - \mathrm{Exc}(\pi)^{(1)})) \simeq \mathscr{O}_X(jK_X)$$

with the same notation as above.

7.15. By Lemma 7.11 and (7.3), the plurigenera $h^0(X, mj_X K_X)$ of a projective variety X with canonical singularities and index j_X are the same as the corresponding plurigenera of any desingularization. If the **k**-algebra

$$R(X, j_X K_X) = \bigoplus_{m \geq 0} H^0(X, mj_X K_X)$$

is finitely generated, the variety $\mathrm{Proj}(R(X, j_X K_X))$ is isomorphic to the canonical model of any desingularization of X.

Examples 7.16 (1) Let X be the quotient of the affine space $\mathbf{A}^n$ by the cyclic group of s-th roots of unity acting by

$$\rho \cdot (x_1, \ldots, x_n) = (\rho x_1, \ldots, \rho x_n)$$

The s-th Veronese map

$$(x_1, \ldots, x_n) \mapsto (\ldots, \prod_{a_1 + \cdots + a_n = s} x_1^{a_1} \ldots x_n^{a_n}, \ldots)$$

induces an isomorphism between X and the affine cone over the s-fold embedding P of $\mathbf{P}^{n-1}$. The blow-up $\pi : Y \to X$ of the origin is smooth, with exceptional divisor E isomorphic to P. The variety Y is also the total space of the line bundle associated with the invertible sheaf $\mathscr{O}_P(-1)$. Write

$$K_Y \sim \pi^* K_X + aE$$

for some rational number a. Since the normal bundle $\mathscr{O}_E(E)$ is isomorphic to $\mathscr{O}_P(-1)$, which is $\mathscr{O}_{\mathbf{P}^{n-1}}(-s)$, and the canonical divisor is $\mathscr{O}_E(K_Y + E)$, we get by restricting to E

$$-n = (a+1)(-s) \qquad \text{hence} \quad a = \frac{n}{s} - 1$$

The variety X *has canonical singularities when* $n \geq s$, *terminal singularities when* $n > s$. One can check ([Re], p. 278) that its index is $s/\mathrm{g.c.d.}(s, n)$.

We have already encountered this singularity in 6.20 with $s = 2$ and $n = 3$, where we showed that it is terminal (see also Section 5.10).

The same construction can be made starting from a smooth subvariety P of a projective space whose canonical sheaf is numerically equivalent to $\mathcal{O}_P(k)$ for some rational number k, and the affine cone X over P. If $\pi : Y \to X$ is the blow-up of the origin, with exceptional divisor E, we get by the same method

$$K_Y \sim \pi^* K_X - (k+1)E$$

For example, a homogeneous hypersurface of degree d in $\mathbf{A}^n$ whose only singular point is the origin has canonical singularities if and only if $n > d$.

(2) **The Reid–Tai criterion.** Consider more generally the quotient X of $\mathbf{A}^n$ by the action

$$\rho \cdot (x_1, \dots, x_n) = (\rho^{\alpha_1} x_1, \dots, \rho^{\alpha_n} x_n)$$

of the group of s-th roots of unity, where ρ is a primitive root. For any integer m, we have

$$\rho^m \cdot (x_1, \dots, x_n) = (\rho^{[m\alpha_1]_s} x_1, \dots, \rho^{[m\alpha_n]_s} x_n)$$

where $[p]_s \in \{0, 1, \dots, s-1\}$ is the remainder of the division of p by s.

If $\sum_{i=1}^{n} [m\alpha_i]_s \geq s$ for all m prime to s, the variety X has canonical singularities ([Re], th. (3.1)). The index is

$$\frac{s}{\text{g.c.d.}(s, \sum_{i=1}^{n} \alpha_i)}$$

([Re], Lemma (3.7)).

For example, the quotient of $\mathbf{k}^3$ for the action

$$\rho \cdot (x_1, x_2, x_3) = (\rho x_1, \rho x_2, \rho^{s-1} x_3)$$

is canonical because $m + m + (s - m) \geq s$ for all m between 0 and $s - 1$. Its index is s. In particular, in any given dimension, there are varieties with canonical singularities and arbitrarily high index. These singularities occur as the only singularities of a general hypersurface of degree $ds(s-1)$ in the weighted projective space $\mathbf{P}(1, 1, s-1, s, s)$.

Here are a few properties of canonical and terminal singularities.

Remarks 7.17 (1) In dimension 2, terminal singularities are smooth, and canonical singularities are rational double points ([CKM], prop. (6.10)).

(2) A general hyperplane section of a projective variety X with canonical (resp. terminal) singularities has canonical (resp. terminal) singularities, and so does the general fiber of a morphism $X \to Y$ ([Re], th. (1.13) and

[CKM], Lemma (6.6)). In particular, a variety with terminal singularities is smooth in codimension 2.

(3) Canonical singularities are rational: if $\pi : Y \to X$ is a resolution of singularities, one has

$$R^i \pi_* \mathscr{O}_Y = 0 \qquad \text{for } i > 0$$

This result is due to Elkik ([El]) and Flenner ([F]). Another proof can be found in [KM], th. 5.22. It uses the Kawamata–Viehweg vanishing theorem 7.21.

As mentioned in 7.7, minimal models are required to have terminal singularities. We will now give two reasons for this requirement.

7.18. First reason. The first reason is that with this property, *minimal models are isomorphic in codimension* 1 (in dimension at least 3, they are, however, not unique; see Exercise 6.7.3). This we will deduce from the following lemma.

Lemma 7.19 *Let X and Y be complex projective varieties nonsingular in codimension* 1, *let $\pi : X \to Y$ be a birational morphism, and let $-D$ be an π-nef[5] Q-Cartier Q-divisor on X. If $\pi_* D$ is effective, so is D.*

If $D = \sum D_k$, where the image by π of each component of D_k has codimension k in Y, the hypothesis is that D_1 is effective.

PROOF OF THE LEMMA. We may assume that D is a Cartier divisor. Since D is effective if and only if a pullback is, we may further assume by Hironaka's resolution of singularities (see 7.22) that X is smooth.

Write $D = \sum D_k$ as above, with D_1 effective.

We proceed by induction on the dimension of X, beginning with the case where X is a surface. Write $D_2 = D_2' - D_2''$, where D_2' and D_2'' are effective divisors without common components, and are contracted by π. For every irreducible curve C contracted by π, we have

$$D_2 \cdot C \le (D_1 + D_2) \cdot C \le 0$$

by hypothesis, hence

$$(D_2'')^2 = D_2' \cdot D_2'' - D_2 \cdot D_2'' \ge 0$$

On the other hand, if H is an ample divisor on Y, we have $D_2'' \cdot \pi^* H = 0$ and $(\pi^* H)^2 > 0$. It follows from the Hodge index theorem ([H1], V, rem. 1.9.1) that D_2'' is numerically trivial, hence trivial since it is effective. It follows that D is effective.

---

[5]This means that D has nonpositive intersection with every curve contracted by π.

Assume now that X has dimension $n > 2$. Let Y' be the intersection of $n - 2$ general hyperplane sections in Y. Both Y' and $X' = \pi^{-1}(Y')$ are surfaces, Y' has isolated singularities, and X' is smooth by Bertini's theorem ([H1], III, cor. 10.9). The divisor $D|_{X'} = (D_1)|_{X'} + (D_2)|_{X'}$ is $\pi|_{X'}$-nef, hence it is effective by induction. Since X' meets every component of D_2, it follows that D_2 is effective.

Let now H be a general (smooth) hyperplane section in X. The morphism $\pi|_H$ factors through the normalization ν of $\pi(H)$ as $\nu \circ \pi_H$. The part of $D|_H$ on which π_H is generically finite is now $(D_1)|_H + (D_2)|_H$, which we have just proved is effective. Since $-D|_H$ is π_H-nef, it follows by induction that $D|_H$ is effective, hence so is D. $\qquad\square$

If X and X' are birationally isomorphic complex minimal models, there exist a smooth projective variety Y with birational morphisms $\pi : Y \to X$ and $\pi' : Y \to X'$, and effective $\mathbf{Q}$-divisors F and F' such that

$$K_Y \sim \pi^* K_X + F \sim \pi'^* K_{X'} + F'$$

Set $D = F' - F$. For any curve C contracted by π, we have

$$D \cdot C = (\pi^* K_X - \pi'^* K_{X'}) \cdot C = -K_{X'} \cdot \pi'_* C \leq 0$$

because $K_{X'}$ is nef. Since $\pi_* D = \pi_* F'$ is effective, the lemma implies that D is effective, hence $F' \geq F$. Since X has terminal singularities, any π-exceptional divisor appears in F, hence in F'. It is therefore π'-exceptional. It follows that $\pi'(\mathrm{Exc}(\pi))$ has codimension at least 2. Similarly, $\pi(\mathrm{Exc}(\pi'))$ has codimension at least 2, and $X - \pi(\mathrm{Exc}(\pi) \cup \mathrm{Exc}(\pi'))$ and $X' - \pi'(\mathrm{Exc}(\pi) \cup \mathrm{Exc}(\pi'))$ are isomorphic.

7.20. Second reason. We have already seen that some kind of singularities must be allowed in Mori's program. For example, a divisorial contraction on a smooth projective variety may lead to terminal nonsmooth singularities (as shown by the contraction $X_1 \to X$ in 6.20). Also, there exist smooth nonuniruled projective varieties that are not birational to any smooth projective variety with nef canonical divisor.[6]

Of course, the class of singularities that we allow must be stable by flips and divisorial contractions, and terminal singularities will been shown to have this property (Proposition 7.44).

7.3   The Kawamata–Viehweg Vanishing Theorem

Cohomological methods are based on the vanishing (of higher cohomology groups) and nonvanishing (of the space of sections) for certain line bundles

---

[6]This is the case for any desingularization of the quotient X of an abelian variety of dimension 3 by the involution $x \mapsto -x$ ([U1], 16.17). Of course, the minimal model here is X itself. It has (locally analytically) the same singularities as above.

on *smooth* projective varieties. These theorems hold only in characteristic zero, and so does Hironaka's resolution of singularities, which we will also use constantly. For simplicity, we will henceforth assume that the base field is $\mathbf{C}$, although all the statements are valid over any algebraically closed field of characteristic zero.

The prototype is Kodaira's vanishing theorem

$$H^i(X, K_X + D) = 0 \qquad \text{for } i > 0$$

when D is an ample divisor on a smooth complex projective variety X. A much finer version is the theorem below due to Kawamata and Viehweg, for whose statement we need a bit of terminology.

If $D = \sum a_i D_i$ is an $\mathbf{R}$-divisor, we define its *round-up* by

$$\lceil D \rceil = \sum \lceil a_i \rceil D_i$$

where $\lceil a_i \rceil$ is the smallest integer greater than or equal to a_i. Its *fractional part* is the effective $\mathbf{R}$-divisor $\lceil D \rceil - D$.

An effective divisor D on a smooth variety X has *simple normal crossings* if for any point x in X, there exist a regular system of parameters $z_1, \ldots, z_n$ at x and a nonnegative integer $r \leq n$ such that $\mathcal{O}_{D_{\text{red}}, x} \simeq \mathcal{O}_{X,x}/(z_1 \cdots z_r)$. In other words, the components of D_{red} are smooth and they intersect transversally. An effective $\mathbf{Q}$-divisor has simple normal crossings if some multiple with integral coefficients has this property.

If D is a divisor with simple normal crossings on X and $\pi : Y \to X$ is a birational projective morphism whose exceptional locus has simple normal crossings, $\pi^* D$ still has simple normal crossings.

Theorem 7.21 (Kawamata–Viehweg vanishing theorem) *Let X be a smooth complex projective variety and let D be a nef and big $\mathbf{Q}$-divisor on X whose fractional part has simple normal crossings. We have*

$$H^i(X, K_X + \lceil D \rceil) = 0$$

for all $i > 0$.

The subtlety here is that numerical properties of the $\mathbf{Q}$-divisor D give information on $\lceil D \rceil$. We will not prove this result here, and refer instead to the existing literature.[7]

---

[7]The original arguments of [Ka3] and [V] used covering constructions to deduce the theorem from a vanishing result for integer divisors. The normal crossing hypothesis is used to control the singularities introduced upon passing to a covering. A number of direct proofs have since been given, one based on connections with logarithmic singularities [EV], another on Hodge theory for twisted coefficient systems [K5], and a third involving singular metrics on line bundles [De2]. We also refer to [CKM], chap. 8, [K4], chap. 9 and 10, and [L].

In Section 7.5, we will see what happens to this theorem when the **Q**-divisor D does not have simple normal crossings. We will also generalize this theorem in different directions in Section 7.4 and in Exercises 7.13.2, 7.13.3, and 7.13.4.

7.22. In the vanishing theorem, there is the annoying (but necessary) hypothesis that the fractional part have normal crossings. This is usually achieved by making use of Hironaka's desingularization theorem in the following form: *let X be a complex variety and let Z be a subscheme of X. There exist a smooth variety Y and a projective morphism $\pi : Y \to X$ such that $\mathrm{Exc}(\pi) \cup \pi^{-1}(Z)$ is a divisor with simple normal crossings.*[8]

7.23. Here is an example of how to use Hironaka's theorem. Let D be an effective Cartier divisor on a variety X. There is a desingularization $\pi : Y \to X$ such that the base-locus of the linear system $|\pi^*D|$ is a divisor with simple normal crossings. This can be proved as follows: first, if X' is the normalization of the graph of the rational map $\varphi_D : X \dashrightarrow \mathbf{P}(H^0(X, D))$ associated with the linear system $|D|$, with projection $\varepsilon : X' \to X$, we can write $\varepsilon^*D = M + F$, where $|M|$ is base-point-free and F is a Cartier divisor that is the base-locus of $|\varepsilon^*D|$. Let $\pi' : Y \to X'$ be a desingularization such that π'^*F has simple normal crossings. By Lemma 7.11,

$$|\pi'^*(\varepsilon^*D)| = \pi'^*|\varepsilon^*D| = \pi'^*|M| + \pi'^*F$$

We may take for π the composite morphism $\varepsilon \circ \pi'$.

7.4   Singularities of Pairs

The reader will probably agree that the Kawamata–Viehweg theorem as stated above is quite unappealing and unnatural. Moreover, it deals exclusively with smooth varieties. We will introduce in this section a formalism well-suited to this theorem, which is much more than just fancy terminology.

The idea is that instead of working with a single variety X, one should consider pairs consisting of X and an effective **Q**-divisor Δ on X such that $[\Delta] = 0$ (i.e., all its coefficients are in $[0, 1)$). This is what is called

---

[8] There are now simple proofs of this statement (see [AJ], [BP]). Hironaka's result is much stronger: it states that there exists a π that is a composition of blow-ups along smooth centers and moreover induces an isomorphism between $X_{\mathrm{reg}} - Z$ and its inverse image (see [Hi1] for the full proofs if you absolutely need to, or [Hi2] for a summary). It implies in particular that if $\pi : Y \to X$ is a birational morphism between smooth complex projective varieties, there exists a composition $Y' \to X$ of blow-ups along smooth centers which factors through π. It is a recent result of Abramovich, Karu, Matsuki and Włodarczyk ([AKMW], [Wl]) that π is a composite of blow-ups and blow-downs with smooth centers.

the *logarithmic* framework. The pair (X, Δ) must satisfy some further requirements which we explain below. The role of K_X is now played by $K_X + \Delta$, and this added flexibility will turn out to be crucial in many applications (see Exercise 7.13.8).

Let $\pi : Y \to X$ be a desingularization and let Δ' be the strict transform of Δ. Assume that $j(K_X + \Delta)$ is a Cartier divisor for some positive integer j. The two lines bundles $\mathscr{O}_Y(jK_Y + j\Delta')$ and $\pi^*\mathscr{O}_X(j(K_X + \Delta))$ are isomorphic over $Y - \text{Exc}(\pi)$. There exists therefore a π-exceptional $\mathbf{Q}$-divisor E on Y such that

$$jK_Y + j\Delta' \equiv \pi^*(j(K_X + \Delta)) + jE \tag{7.4}$$

and E is uniquely defined, by 7.12. When we write

$$K_Y + \Delta' \sim \pi^*(K_X + \Delta) + E \tag{7.5}$$

it will always be understood that the divisor E is the one defined by (7.4).

Definition 7.24 *Let X be a normal projective variety and let Δ be an effective $\mathbf{Q}$-divisor on X with $[\Delta] = 0$. The pair (X, Δ) is log terminal[9] if $K_X + \Delta$ is a $\mathbf{Q}$-Cartier $\mathbf{Q}$-divisor and there exists a desingularization $\pi : Y \to X$ such that, with the notation above, $\text{Exc}(\pi) + \Delta'$ is a divisor[10] with simple normal crossings, and $\lceil E \rceil \geq 0$ in (7.5).*

As in 7.14, the *index* $j_{X,\Delta}$ of the pair (X, Δ) is the smallest positive integer j such that $j(K_X + \Delta)$ is a Cartier divisor.

The reader will find in [KM], def. 2.34, a much more conceptual definition which does not involve the choice of a desingularization (note also that the effectivity of Δ is sometimes not part of the definition).

Remarks 7.25 (1) If a projective variety X has canonical singularities, the pair $(X, 0)$ is log terminal.

(2) The pair consisting of a smooth projective variety and an effective $\mathbf{Q}$-divisor with simple normal crossings, all of whose coefficients are < 1, is log terminal (take $Y = X$ in the definition).

If the pair (X, Δ) is log terminal, the defining property $\lceil E \rceil \geq 0$ will hold for *any* desingularization $\pi : Y \to X$.

Indeed, since any two desingularizations can be dominated by a third, this is equivalent to proving that if Δ is an effective $\mathbf{Q}$-divisor with simple normal crossings on a smooth variety X such that $[\Delta] = 0$, for *any*

---

[9]This is usually called *Kawamata log terminal*, or klt, in the literature.

[10]Since we are not assuming that X is locally $\mathbf{Q}$-factorial, the exceptional locus of a resolution may have components with codimension more than 1. This creates some technical problems and gives rise to many other definitions for singularities of pairs; see, for example, pp. 58–61 in [KM].

desingularization $\pi : Y \to X$ as in the definition, we have $\lceil E \rceil = 0$ (with the same notation). By the strong version of Hironaka's theorem (see footnote 8, p. 184), any desingularization can be dominated by a composition of smooth blow-ups. It is therefore enough to check the property $\lceil E \rceil \geq 0$ when π is the blow-up of a smooth codimension c subvariety Z of X, with exceptional divisor F. Write $\Delta = \sum_i n_i D_i$. Since Δ has simple normal crossings, the set $I = \{i \mid Z \subset D_i\}$ has at most c elements, and

$$K_Y \equiv \pi^* K_X + (c-1)F \qquad \text{and} \qquad \pi^* \Delta = \Delta' + \sum_{i \in I} n_i F$$

hence

$$E = \left(c - 1 - \sum_{i \in I} n_i \right) F$$

and $\lceil E \rceil$ is effective because $0 \leq n_i < 1$ for all i.

With this new formalism, the Kawamata–Viehweg vanishing theorem (Theorem 7.21) takes a much more natural form: it is simply the analog of the usual Kodaira vanishing theorem where K_X has been replaced with $K_X + \Delta$.

Theorem 7.26 (Logarithmic Kawamata–Viehweg vanishing theorem) *Let (X, Δ) be a projective log terminal pair and let D be a nef and big $\mathbf{Q}$-Cartier $\mathbf{Q}$-divisor on X such that $K_X + \Delta + D$ is a Cartier divisor.[11] We have*

$$H^i(X, K_X + \Delta + D) = 0$$

for all $i > 0$.

We show in Exercises 7.13.2 and 7.13.3 how to deduce this theorem from the standard Kawamata–Viehweg vanishing theorem 7.21. In Exercise 7.13.4, we give a relative version of Theorem 7.26. The reader will notice that the definition of log terminal is exactly what one needs to make the proofs work.

We will actually use this theorem very seldom: in most applications, one needs to work on a blow-up of X that might as well be taken smooth, in which case the logarithmic version of the Kawamata–Viehweg theorem reduces to the standard version.

7.5   Multiplier Ideal of an Effective Q-divisor

There is another direction in which the vanishing theorems have been generalized. Instead of trying to prove a Kodaira type result with the weakest

---

[11]What we mean here is that there is a Cartier divisor M which is numerically equivalent to $K_X + \Delta + D$, and the conclusion is $H^i(X, M) = 0$ for $i > 0$.

possible hypotheses (in the spirit of Theorem 7.26), a different approach coming from analytic geometry consists instead of proving a general vanishing theorem with a "correction term" that measures in a sense how far the divisor is from having simple normal crossings (or, in the logarithmic setting, how far the pair is from being log terminal). This "correction term" takes the form of an ideal called the *multiplier ideal*. Because of the importance that this theory has recently taken and the results it has achieved, I will briefly introduce it, although it will not be used in this book. I refer to the forthcoming book by R. Lazarsfeld ([L]) for further developments.

Let D be an effective **Q**-divisor on a *smooth* complex variety X. By Hironaka's theorem, there exist a smooth variety Y and a projective morphism $\pi : Y \to X$ such that $\mathrm{Exc}(\pi) + \pi^* D$ is a divisor with simple normal crossings. Define

$$\mathscr{I}_D = \pi_* \mathscr{O}_Y (K_Y - \pi^* K_X - [\pi^* D]) \qquad (7.6)$$

Using (1.11) and Lemma 7.11, we get

$$\mathscr{I}_D \subset \pi_* \mathscr{O}_Y (\mathrm{Ram}(\pi)) \simeq \mathscr{O}_X$$

so that $\mathscr{I}_D$ is indeed a sheaf of ideals whose cosupport is contained in the support of D. By 7.12, this ideal is trivial if and only if every exceptional divisor of π appears with multiplicity > -1 in $K_Y - \pi^* K_X - [\pi^* D]$. This is exactly saying that the pair (X, D) is *log terminal* in the sense of Definition 7.24.

The projection formula implies

$$\mathscr{I}_D = \mathscr{I}_{D-[D]}(-[D])$$

If D has simple normal crossings, Remark 7.25(2) implies that the pair $(X, D - [D])$ is log terminal, hence

$$\mathscr{I}_D = \mathscr{O}_X(-[D]) \qquad (7.7)$$

Finally, one needs to check that this ideal $\mathscr{I}_D$ does not depend on the choice of the resolution π. Since any two resolutions can be dominated by a third, this follows from (7.7). The ideal sheaf $\mathscr{I}_D$ is called the *multiplier ideal* associated with D.

Example 7.27 Let C be the curve $y^2 = x^3$ in $\mathbf{C}^2$. Consider the composition $\pi : Y \to \mathbf{C}^2$ of blow-ups

We have

$$\begin{aligned}\pi^* C &= C + 2E_1 + 3E_2 + 6E_3 \\ K_Y &\sim E_1 + 2E_2 + 4E_3\end{aligned}$$

hence, for any positive rational number a,

$$K_{Y/X} - [\pi^*(aC)] \sim -[a]C + (1 - [2a])E_1 + (2 - [3a])E_2 + (4 - [6a])E_3$$

It follows that $\mathscr{I}_{aC}$ is trivial if and only if $[6a] \le 4$, i.e., $a < \frac{5}{6}$. Moreover,

$$\mathscr{I}_{\frac{5}{6}C} = \pi_* \mathscr{O}_Y(E_3) = (x, y)$$

The vanishing theorem 7.21 now takes the following more general form.

Theorem 7.28 (Nadel vanishing theorem) *Let X be a smooth complex projective variety and let D be a nef and big $\mathbf{Q}$-divisor on X. We have*

$$H^i(X, \mathscr{I}_{[D]-D}(K_X + [D])) = 0$$

for all $i > 0$.

If the fractional part of D has simple normal crossings, we recover Theorem 7.21. If the pair $(X, [D] - D)$ is log terminal, we recover Theorem 7.26 in the case where X is smooth. There is also a logarithmic version of Nadel's vanishing theorem, which generalizes all these results, where the line bundle $\mathscr{O}_X(K_X + \Delta + D)$ is twisted by an ideal whose cosupport is the locus of points where the pair (X, Δ) is *not* log terminal ([K7], (2.16.1)).

Exercise 7.13.7 presents a nice application (due to Kollár) of the vanishing theorem 7.28 to singularities of a theta divisor Θ on a principally polarized abelian variety A: it is shown that for any positive integer k, the locus of points of A that have multiplicity k on Θ has codimension at least k in A. For $k = 1$, this result is empty. For $k = 2$, it says that Θ is reduced, a classical result. For $k = 3$, it says that if Θ is not normal, a general point of a component of $\mathrm{Sing}(\Theta)$ of maximal dimension has multiplicity 2. Using analogous methods but yet another vanishing theorem, Ein and Lazarsfeld have proved in [EL] that in this case, Θ must be reducible. It is then classical that the principally polarized abelian variety (A, Θ) decomposes as a product of lower-dimensional principally polarized abelian varieties.

7.6   The Nonvanishing Theorem

The following theorem is due to Shokurov ([Sh1]). It gives sufficient conditions for a linear system to be nonempty and is a first step toward the base-point-free theorem 7.32. Since the proofs of both theorems are quite technical but involve similar ideas that are more transparent in the proof of the base-point-free theorem 7.32, we will first explain the common basic underlying strategy before plunging into technicalities.

Assume for simplicity that we are on a smooth projective variety X. Given a nef divisor D on X, we need to produce elements of $|mD|$ for m large, knowing say that $aD - K_X$ is ample. Assume we can write, for some integer m,

$$mD = K_X + H + F$$

where H is ample and F is a smooth hypersurface on X. By the Kodaira vanishing theorem, there is a surjection

$$H^0(X, K_X + H + F) \twoheadrightarrow H^0(F, K_F + H)$$

and we can hope to get sections by induction on the dimension. This is, however, too strong a hypothesis. We can use a blow-up $\pi : Y \to X$ to get more elbow room: by Lemma 7.11, for any effective π-exceptional divisor E, we have

$$H^0(X, mD) \simeq H^0(X, m\pi^*D + E)$$

and we need to have vanishing for

$$H^1(Y, m\pi^*D + E - F)$$

In other words, we need to write

$$m\pi^*D \sim K_Y + \Delta_Y + (\text{nef and big}) - (\text{effective exceptional}) + F$$

Now write

$$K_Y + \pi^*A \sim \pi^*K_X + \sum_i a_i F_i$$

and assume that we know (this is the content of the nonvanishing theorem) that some $\mathcal{O}_X(bD)$ has a nonzero section. We choose π such that

$$b\pi^*D \equiv M + \sum_i r_i F_i$$

where $|M|$ is base-point-free (possibly 0). We have

$$m\pi^*D \sim K_Y + (m - cb - a)\pi^*D + cM + \pi^*(aD - K_X) - \sum_i (a_i - cr_i)F_i$$

Assume $m > cb + a$ and $c \geq 0$, so that

$$(m - cb - a)\pi^*D + cM + \pi^*(aD - K_X)$$

is nef and big. We need to choose $c \geq 0$ in such a way that

$$\sum_i (a_i - cr_i)F_j = (\text{effective exceptional}) - F - \Delta_Y$$

which we do by arranging $\min_i(a_i - cr_i) = -1$ (it is possible to perturb the coefficients in such a way that the minimum is obtained for only one i). We see here how the condition $a_i > -1$, which defines log terminality, comes into play.

This is the main idea that underlies the proofs of both the nonvanishing theorem and the base-point-free theorem.

Before proceeding with these proofs, we will first put behind us the proof of the following result, very much in the spirit of Corollary 1.32.

Lemma 7.29 *Let X be a complex projective variety and let M be a nef and big $\mathbf{Q}$-divisor on X. There exist a desingularization $\pi : Y \to X$ and a reduced divisor $\sum F_i$ on Y with simple normal crossings that contains $\mathrm{Exc}(\pi)$, such that, for any $\eta > 0$, there exist rational numbers $p_i \in (0, \eta)$ such that $\pi^* M - \sum p_i F_i$ is ample.*

PROOF. Since the inverse image of a nef and big divisor by a generically finite morphism has the same properties, we may assume by 7.22 that X is smooth. By Corollary 1.32, there exists an effective $\mathbf{Q}$-divisor D on X such that $M - tD$ is ample for all rationals $t \in (0,1)$. By 7.22, there is a desingularization $\pi : Y \to X$ such that $\mathrm{Supp}(\pi^* \lceil D \rceil) \cup \mathrm{Exc}(\pi)$ is a divisor $\sum F_i$ with simple normal crossings. Write $\pi^* D = \sum d_i F_i$. By 1.42, there exist nonnegative integers $a_i(t)$ and $m(t)$ such that

$$\pi^*(M - tD) - \sum \frac{a_i(t)}{m} F_i = \pi^* M - \sum \left(\frac{a_i(t)}{m} + td_i \right) F_i$$

is ample for all $t \in (0,1)$ and $m \geq m(t)$.   □

Now comes the main theorem of this section. Recall that the "boundary divisor" Δ that appears in a log terminal pair (X, Δ) is a $\mathbf{Q}$-divisor with coefficients in $[0, 1)$ such that $K_X + \Delta$ is a $\mathbf{Q}$-Cartier divisor. The pair consisting of a projective variety with canonical singularities and the zero divisor is log terminal.

Theorem 7.30 (Nonvanishing theorem) *Let (X, Δ) be a complex projective log terminal pair, let D be a nef Cartier divisor on X, and let A be an effective Cartier divisor on X such that $aD + A - (K_X + \Delta)$ is nef and big for some positive rational a. We have*

$$H^0(X, mD + A) \neq 0$$

for all integers m sufficiently large.

This theorem is often used with $A = 0$. However, the presence of A adds just the extra flexibility needed in the proof (this is apparent in the induction argument at its very end) and in applications (such as the very end of the proof of the base-point-free theorem 7.32, p. 197).

7.31. Under the hypotheses of the theorem, the vanishing theorem 7.26 gives $H^i(X, mD + A) = 0$ for all integers $m \geq a$, hence

$$h^0(X, mD + A) = \chi(X, mD + A)$$

Moreover, these numbers are nonzero for all integers m sufficiently large by the theorem. Since $P(m) = \chi(X, mD + A)$ is a polynomial of degree at most $\dim(X)$, it cannot vanish at $\dim(X) + 1$ consecutive integers. It follows that there exists an integer $m \leq \lceil a \rceil + \dim(X)$ that satisfies the conclusion of the theorem.

PROOF OF THE THEOREM. *First step: reduction to the case where D is not numerically trivial.*

If D is numerically trivial, $mD + A - (K_X + \Delta)$ is nef and big for all integers m, and $\chi(X, mD+A) = h^0(X, mD+A)$ by the vanishing theorem. By the Riemann–Roch theorem, this number only depends on the class of $mD + A$, hence not on m, and it is nonzero because A is effective. This proves the theorem in this case.

Second step: reduction to the case where X is smooth and $aD + A - (K_X + \Delta)$ is ample.

Let now $\pi : Y \to X$ be a desingularization such that $\mathrm{Exc}(\pi) + \Delta'$ has simple normal crossings and write as in (7.4)

$$K_Y + \Delta' \sim \pi^*(K_X + \Delta) + E$$

where E is π-exceptional and satisfies $\lceil E \rceil \geq 0$. Write $E = E' - E''$, where E' and E'' are effective without common components. By Lemma 7.29, we may assume, after further desingularizing Y, that

$$
\begin{aligned}
M' &= \pi^*(aD + A - (K_X + \Delta)) - \sum_i p_i F_i \\
&= a\pi^*D + (\pi^*A + \lceil E' \rceil) - \left(K_Y + \Delta' + E'' + \lceil E' \rceil - E' + \sum_i p_i F_i\right)
\end{aligned}
$$

is ample for arbitrarily small positive p_i, which we may choose in such a way that the coefficients of the simple normal crossings effective divisor

$$\Delta'' = \Delta' + E'' + \lceil E' \rceil - E' + \sum_i p_i F_i$$

are all less than 1. The pair (Y, Δ'') is then log terminal. Since the vector spaces

$$H^0(Y, m\pi^*D + \pi^*A + \lceil E' \rceil) \quad \text{and} \quad H^0(X, mD + A)$$

are isomorphic by Lemma 7.11, it is therefore enough to prove the theorem assuming that X is smooth and $aD + A - (K_X + \Delta)$ is ample.

Pick a point x_0 in $X - \text{Supp}(A + \Delta)$.

Third step: there exist positive integers b and t and a divisor $M \sim t(bD + A - (K_X + \Delta))$ with multiplicity $> 2t \dim(X)$ at x_0.

Let n be the dimension of X. Since D is nef but not numerically trivial, there exists an irreducible curve C such that $D \cdot C > 0$. For any ample divisor H on X, there exist a positive integer m and hypersurfaces $H_1, \ldots, H_{n-1}$ in $|mH|$ containing C whose intersection has pure dimension 1. Since D is nef, we have

$$D \cdot (mH)^{n-1} = D \cdot H_1 \cdots \cdots H_{n-1} \geq D \cdot C > 0$$

hence $D \cdot H^{n-1}$ is positive. Since D is nef, we also have $D^i \cdot H^{n-i} \geq 0$ for all $i > 0$ (Lemma 1.23). In particular, for $b \geq a$, the binomial expansion of $((b-a)D + aD + A - (K_X + \Delta))^n$ yields

$$
\begin{aligned}
(bD + A - (K_X + \Delta))^n &\geq n(b-a)D \cdot (aD + A - (K_X + \Delta))^{n-1} \\
&= \alpha(b-a)
\end{aligned}
$$

where α is positive. Let j be an integer such that $j\Delta$ has integral coefficients. As in the proof of Proposition 5.17, it follows that there exist a positive integer t divisible by j and a divisor M in $|t(bD + A - (K_X + \Delta))|$ such that

$$
\begin{aligned}
\text{mult}_{x_0} M &\geq \tfrac{1}{2}t \sqrt[n]{(bD + A - K_X)^n} \\
&\geq \tfrac{1}{2}t \sqrt[n]{\alpha(b-a)} \\
&\geq 2tn
\end{aligned}
$$

for b sufficiently large.

Fourth step: construction of a blow-up $\pi : Y \to X$ such that $\pi^*(mD + A)$ is numerically equivalent to

$$K_Y + \Delta_Y + (\text{ample}) + (\text{effective}) - (\text{effective exceptional}) + (\text{smooth})$$

Let $\varepsilon : \tilde{X} \to X$ be the blow-up of x_0 and let $\pi : Y \to X$ be the desingularization obtained by applying Lemma 7.29 to $\tilde{X}$ and the nef and big divisor $\varepsilon^*(aD + A - (K_X + \Delta))$. Let F_0 be the exceptional divisor in Y that dominates $\varepsilon^{-1}(x_0)$. Write

(a) $K_Y + \pi^*A \sim \pi^*(K_X + \Delta) + \sum_i a_i F_i$, with $a_i > -1$ for all i (because A is effective and the pair (X, Δ) is log terminal), and $a_0 = n - 1$ (because x_0 is not in the support of A and Δ);

(b) $\pi^*M = \sum_i r_i F_i$, with $r_i \geq 0$ and $r_0 \geq 2tn$ by construction;

where the F_i are divisors on Y (not necessarily exceptional).

By Lemma 7.29, there exist $p_i \in (0, a_i + 1)$ such that

$$\pi^*(aD + A - (K_X + \Delta)) - \sum_i 2p_i F_i$$

is ample.

For any positive integer m and positive rational c, set

$$
\begin{aligned}
N_{m,c} &= m\pi^*D - K_Y + \sum_i (-cr_i + a_i - p_i)F_i \\
&\sim m\pi^*D + \pi^*A - \pi^*(K_X + \Delta) - c\pi^*M - \sum_i p_i F_i \\
&\sim m\pi^*D + \pi^*(A - (K_X + \Delta)) - ct\pi^*((b-a)D) \\
&\qquad - ct\pi^*(aD + A - (K_X + \Delta)) - \sum_i p_i F_i \\
&\sim (m - a - ct(b-a))\pi^*D \\
&\qquad + (1 - ct)\pi^*(aD + A - (K_X + \Delta)) - \sum_i p_i F_i \\
&\sim (m - a - ct(b-a))\pi^*D + \left(\tfrac{1}{2} - ct\right)\pi^*(aD + A - (K_X + \Delta)) \\
&\qquad + \tfrac{1}{2}\left(\pi^*(aD + A - (K_X + \Delta)) - \sum_i 2p_i F_i\right)
\end{aligned}
$$

For $ct \le \tfrac{1}{2}$ and $m \ge \tfrac{1}{2}(a+b)$, the first two terms in the sum of the last line are nef and the last one is ample, hence $N_{m,c}$ is ample by 1.22.

Since $a_i - p_i > -1$ for all i and the r_i are not all zero,

$$c = \min_i \frac{1 + a_i - p_i}{r_i}$$

is a positive rational and $\min_i(-cr_i + a_i - p_i) = -1$. Furthermore, after wiggling the p_i a little, *the minimum is obtained for only one i*, say $i = i_0$. Let F be the (smooth) corresponding F_{i_0}. Note that

$$c \le \frac{1 + a_0 - p_0}{r_0} < \frac{n}{2tn}$$

hence $ct \le \tfrac{1}{2}$. With our choice of c, the divisor $N_{m,c}$ is therefore ample for all $m \ge \tfrac{1}{2}(a+b)$. We have

$$K_Y + N_{m,c} = m\pi^*D + \sum_i(-cr_i + a_i - p_i)F_i = m\pi^*D + B - F$$

where $\lceil B \rceil$ is an effective divisor with simple normal crossings in which F does not appear. Since $-cr_i + a_i - p_i \le a_i$ and $b_{i_0} > -1$, we get $\lceil B \rceil \le \sum_i \lceil a_i \rceil F_i$, hence, using (a),

$$\pi_* \lceil B \rceil \le \sum_i \lceil a_i \rceil \pi_* F_i = A$$

from which follows $\lceil B \rceil \leq \pi^* A + \lceil B \rceil_{\text{exc}}$, where $\lceil B \rceil_{\text{exc}}$ is the exceptional part of $\lceil B \rceil$. In short, $\pi^*(mD + A)$ is numerically equivalent to

$$K_Y + (\lceil B \rceil - B) + N_{m,c} + (\pi^* A + \lceil B \rceil_{\text{exc}} - \lceil B \rceil) - \lceil B \rceil_{\text{exc}} + F$$

which is of the desired form

$$K_Y + \Delta_Y + (\text{ample}) + (\text{effective}) - (\text{effective exceptional}) + (\text{smooth})$$

Fifth step: applying the Kawamata–Viehweg vanishing theorem.

By the remarks made at the beginning of this section, it is enough to prove that the divisor

$$\pi^*(mD + A) - (\pi^* A + \lceil B \rceil_{\text{exc}} - \lceil B \rceil) + \lceil B \rceil_{\text{exc}} = m\pi^* D + \lceil B \rceil$$

which is numerically equivalent to

$$K_Y + (\lceil B \rceil - B) + N_{m,c} + F = K_Y + \lceil N_{m,c} \rceil + F$$

has nonzero sections. By the vanishing theorem 7.21, we have

$$H^1(Y, K_Y + \lceil N_{m,c} \rceil) = 0$$

hence the restriction

$$H^0(Y, m\pi^* D + \lceil B \rceil) \to H^0(F, m\pi^* D + \lceil B \rceil)$$

is surjective for all $m \geq \frac{1}{2}(a + b)$. Since

$$\left(\tfrac{1}{2}(a + b)\pi^* D + B \right)|_F - K_F \sim N_{\frac{1}{2}(a+b),c}|_F$$

is ample and $\pi^* D|_F$ is nef, an induction hypothesis shows that the vector space $H^0(F, m\pi^* D + \lceil B \rceil)$ is nonzero for all b sufficiently large, and so is $H^0(Y, m\pi^* D + \lceil B \rceil)$ (here we use the fact that since F does not appear in B, we have $\lceil B|_F \rceil = \lceil B \rceil|_F$).

This finishes the proof of the nonvanishing theorem. $\square$

7.7   The Base-Point-Free Theorem

Using the vanishing and nonvanishing theorems, we will now prove an important result of Kawamata's. For more explanations about the underlying ideas of the proof, see Section 7.6.

Recall again that the "boundary divisor" Δ that appears in a log terminal pair (X, Δ) is a $\mathbf{Q}$-divisor with coefficients in $[0, 1)$ such that $K_X + \Delta$ is a $\mathbf{Q}$-Cartier divisor. The pair consisting of a projective variety with canonical singularities and the zero divisor is log terminal.

Theorem 7.32 (Base-point-free theorem) *Let (X, Δ) be a complex[12] projective log terminal pair and let D be a nef Cartier divisor on X such that $aD - (K_X + \Delta)$ is nef and big for some positive rational a. The linear system $|mD|$ is base-point-free for all integers m sufficiently large.*

The hypothesis "D nef" is obviously necessary. The theorem implies that if X is a projective variety with canonical singularities with K_X nef and big, and $j_X K_X$ is a Cartier divisor, the linear system $|mj_X K_X|$ is base-point-free for all integers m sufficiently large. This is the abundance conjecture for varieties of general type (see 7.7).

Corollary 7.33 *Let X be a complex projective variety with canonical singularities and index j_X. If K_X is nef and big, the canonical algebra*

$$R(X, K_X) = \bigoplus_{m \geq 0} H^0(X, mj_X K_X)$$

is finitely generated.

There are logarithmic versions of these statements: if (X, Δ) is a projective log terminal pair such that $K_X + \Delta$ is nef and big, with index $j_{X,\Delta}$, the linear system $|mj_{X,\Delta}(K_X + \Delta)|$ is base-point-free for all integers m sufficiently large. There is also an analog of the canonical algebra in this situation and it is still finitely generated ([KM], th. 3.11).

The theorem also implies, for a *smooth* projective variety X, that some multiple of a supporting divisor for a K_X-negative extremal ray is base-point-free hence defines the contraction of that ray (see Corollary 6.9). We will come back to that in Section 7.10.

PROOF OF THE THEOREM. The nonvanishing theorem 7.30 implies that $|mD|$ is nonempty for all m sufficiently large.

For any integer $b \geq 2$, let $B(b)$ be the base-locus of $|bD|$. Since X is Noetherian, the nonincreasing sequence $(B(b^r))_r$ stabilizes for r large to a proper closed subset $B_\infty(b)$ of X. If all $B_\infty(b)$ are empty,

$$B(2^r) = B(3^r) = \varnothing$$

for r large enough. Since any integer m large enough is a linear combination with nonnegative coefficients of 2^r and 3^r, we have then $B(m) \subset B(2^r) \cup B(3^r) = \varnothing$ and the theorem is proved.

We will henceforth assume $B_\infty(b) = B(b^r) \neq \varnothing$ for some $b \geq 2$ and derive a contradiction. By 7.23, there is a desingularization $\tau : Y' \to X$ such that

$$b^r \tau^* D \sim M + E$$

---

[12]Using totally different ideas, Keel has recently obtained in [K] (th. (0.5)) a similar base-point-free theorem valid on **Q**-factorial normal threefolds defined over the algebraic closure of a finite field.

where E an effective divisor with simple normal crossings which is the base-locus of $|b^r\tau^*D|$, and $|M|$ is base-point-free. This property will be preserved under any further desingularization whose exceptional locus has simple normal crossings. Applying Lemma 7.29 to the nef and big divisor $\tau^*(aD - (K_X + \Delta))$, we may assume that we have a desingularization $\pi : Y \to X$ and divisors F_i on Y (not necessarily exceptional) such that

(a) the linear system $|b^r\pi^*D - \sum_i r_i F_i|$ is base-point-free, where $r_i \geq 0$ for all i, and $\bigcup_{r_i > 0} \pi(F_i) = B_\infty(b)$;

(b) the divisor $\pi^*(aD - (K_X + \Delta)) - \sum_i p_i F_i$ is ample, where $p_i \in (0,1)$ for all i;

(c) $K_Y \sim \pi^*(K_X + \Delta) + \sum_i a_i F_i$, with $a_i > -1$ for all i (since the pair (X, Δ) is log terminal).

For any positive integer m and positive rational c, set

$$
\begin{aligned}
N_{m,c} &= m\pi^*D - K_Y + \sum_i (-cr_i + a_i - p_i)F_i \\
&\sim (m - cb^r - a)\pi^*D + c\left(b^r\pi^*D - \sum_i r_i F_i\right) \\
&\quad + \left(\pi^*(aD - (K_X + \Delta)) - \sum_i p_i F_i\right)
\end{aligned}
$$

For $m \geq cb^r + a$, the first term of this last sum is nef, the second term is base-point-free, hence nef, and the last term is ample, hence $N_{m,c}$ is ample (1.22).

Because $B_\infty(b) \neq \varnothing$, not all r_i are zero. Since $a_i - p_i > -1$ for all i,

$$
c = \min_i \frac{1 + a_i - p_i}{r_i}
$$

is a positive rational and $\min_i(-cr_i + a_i - p_i) = -1$. Furthermore, after wiggling the p_i a little (this will not affect property (b) above), *the minimum is obtained for only one i*, say $i = i_0$, and $r_{i_0} > 0$. Let F be the (smooth) corresponding F_{i_0}. We have then

$$
K_Y + N_{m,c} = m\pi^*D + B - F
$$

where $\lceil B \rceil$ is an effective divisor in which F does not appear, and which is exceptional because $-cr_i + a_i - p_i$ can only be positive if a_i is.

The vanishing theorem 7.21 implies

$$
H^1(Y, K_Y + \lceil N_{b,c} \rceil) = 0
$$

hence the restriction

$$
H^0(Y, m\pi^*D + \lceil B \rceil) \to H^0(F, m\pi^*D + \lceil B \rceil)
$$

is surjective. Take for m a power of b larger than $cb^r + a$. By Lemma 7.11, the vector space $H^0(Y, m\pi^*D + \lceil B \rceil)$ is isomorphic to $H^0(X, mD)$; hence all its elements vanish on F (because $\pi(F)$ is in $B(m) = B_\infty(b)$, since $r_{i_0} > 0$). It follows that $H^0(F, m\pi^*D + \lceil B \rceil)$ vanishes for all such m. Since

$$\left(N_{\lceil cb^r + a \rceil, c}\right)|_F = (\lceil cb^r + a \rceil \pi^*D + B)|_F - K_F$$

is ample and $\lceil B \rceil|_F$ is effective and has simple normal crossings, this contradicts the nonvanishing theorem 7.30.   □

7.8   The Rationality Theorem

This is the last difficult step on the road to the cone theorem, which is a formal geometrical consequence thereof. We will also state it in its logarithmic version.

Recall that the "boundary divisor" Δ that appears in a log terminal pair (X, Δ) is a **Q**-divisor with coefficients in $[0, 1)$ such that $K_X + \Delta$ is a **Q**-Cartier divisor. The pair consisting of a projective variety with canonical singularities and the zero divisor is log terminal. The index $j_{X,\Delta}$ of a pair (X, Δ) is the smallest integer j such that $j(K_X + \Delta)$ is a Cartier divisor.

Theorem 7.34 *Let (X, Δ) be a complex projective log terminal pair such that $K_X + \Delta$ is not nef. Let H be a nef and big Cartier divisor on X.*

The number $r = \sup\{t \in \mathbf{R} \mid H + t(K_X + \Delta) \text{ nef}\}$ is rational and

$$\frac{r}{j_{X,\Delta}} = \frac{u}{v} \qquad \text{with} \quad 0 < v \le j_{X,\Delta}(\dim(X) + 1)$$

This theorem still holds when H is only nef, with the weaker conclusion

$$\frac{r}{j_{X,\Delta}} = \frac{u}{v} \qquad \text{with} \quad 0 < v \le 2j_{X,\Delta}\dim(X)$$

(see Exercise 7.13.5), where it is obtained as a corollary of Kawamata's result (Corollary 7.48) on lengths of extremal rays.

7.35. Assume H is nef and big (resp. ample). The set

$$\{t \in \mathbf{R}^+ \mid H + t(K_X + \Delta) \text{ nef}\}$$

is the closed interval $[0, r]$, and $H + t(K_X + \Delta)$ is nef and big (resp. ample) when $t \in \mathbf{Q} \cap [0, r)$ (write

$$H + t(K_X + \Delta) = \left(1 - \frac{t}{r}\right)H + \frac{t}{r}(H + r(K_X + \Delta))$$

and use 1.22 (resp. 1.29).

Let me give a rough idea of the proof. Assume for simplicity that X is smooth and $\Delta = 0$. Consider the linear systems

$$|pH + qK_X|$$

If r is irrational, the set of pairs (p, q) of positive integers such that

$$\frac{q-1}{p} < r < \frac{q}{p}$$

is infinite (this is the set Λ_1 of the proof). For such a pair, $pH + (q-1)K_X$ is nef and big, so that we can apply vanishing to $pH + qK_X$, whereas $pH + qK_X$ is not nef, hence has base-points. The base-locus stabilizes for all sufficiently large pairs (second step) and is distinct from X (third step: this follows from the logarithmic vanishing theorem applied to $pH + qK_X$). We then follow the same ideas as in the proof of the base-point-free theorem 7.32: write, on some blow-up $\pi : Y \to X$, the divisor $\pi^*(pH + qK_X)$ as

$$K_Y + \Delta_Y + (\text{ample}) - (\text{effective exceptional}) + (\text{smooth})$$

and use vanishing to prove that all cohomology groups of

$$\pi^*(pH + qK_X) + (\text{effective exceptional})$$

must vanish for too many pairs (p, q) to derive a contradiction.

PROOF OF THE THEOREM. We may assume $r > 0$, since otherwise there is nothing to prove. Let n be the dimension of X.

First step: we may assume that H and $H + j_{X,\Delta}(K_X + \Delta)$ are base-point-free.

Let t be an integer larger than $\frac{2j_{X,\Delta}}{r}$. By 7.35, the divisors

$$tH + j_{X,\Delta}(K_X + \Delta)$$

and

$$tH + j_{X,\Delta}(K_X + \Delta) - (K_X + \Delta)$$

are nef and big, hence

$$H' = m(tH + j_{X,\Delta}(K_X + \Delta))$$

is base-point-free by Theorem 7.32 for all m sufficiently large, and it is still nef and big. Write

$$H' + j_{X,\Delta}(K_X + \Delta) =$$
$$\frac{m-1}{3}(2tH + j_{X,\Delta}(K_X + \Delta)) + \frac{m+2}{3}(tH + 2j_{X,\Delta}(K_X + \Delta))$$

Since $2tH + j_{X,\Delta}(K_X + \Delta)$ and $2tH + j_{X,\Delta}(K_X + \Delta) - (K_X + \Delta)$ are nef and big, and $tH + 2j_{X,\Delta}(K_X + \Delta)$ and $tH + 2j_{X,\Delta}(K_X + \Delta) - (K_X + \Delta)$ are nef and big, each term in this sum is again base-point-free for m sufficiently large $\equiv 1 \pmod{3}$, hence so is $H' + j_{X,\Delta}(K_X + \Delta)$.

Moreover, since

$$H' + r'(K_X + \Delta) = mt\left(H + \frac{r' + mj_{X,\Delta}}{mt}(K_X + \Delta)\right)$$

we get $r = \frac{r' + mj_{X,\Delta}}{mt}$, and r is rational if and only if r' is. Once we know that r is rational, we may choose t and m prime to its denominator v (with $m \equiv 1 \pmod{3}$). This implies that v divides the denominator of r'. It is therefore enough to prove the theorem for H' and r'.

Let η be a positive rational and let Λ_η be the set of pairs of positive integers (p, q) such that

$$\frac{j_{X,\Delta}q - \eta}{p} < r < \frac{j_{X,\Delta}q}{p}$$

or equivalently $0 < j_{X,\Delta}q - rp < \eta$. This set is infinite if r is irrational, infinite or empty otherwise.

Indeed, if r is irrational, this follows from a classical result: for any positive real numbers s and η, there exist arbitrarily large integers p and q such that $0 < q - ps < \eta$. If r is rational and $(p_0, q_0) \in \Lambda_\eta$, simply take $p = p_0 + dj_{X,\Delta}$ and $q = q_0 + dr$ for any positive integer d.

If $(p, q) \in \Lambda_\eta$, note that $pH + (qj_{X,\Delta} - \eta)(K_X + \Delta)$ is nef and big, whereas $pH + qj_{X,\Delta}(K_X + \Delta)$ is not nef, hence $|pH + qj_{X,\Delta}(K_X + \Delta)|$ has base-points. Let $B(p, q)$ be its base-locus.

Second step: if $(p_0, q_0) \in \Lambda_1$, we have $B(p, q) \subset B(p_0, q_0)$ for all $(p, q) \in \Lambda_1$ with q sufficiently large. Since X is Noetherian, there exists therefore a closed subset B_∞ of X such that $B(p, q) = B_\infty$ for all $(p, q) \in \Lambda_1$ with q sufficiently large.

Let $(p, q) \in \Lambda_1$ and let $q = \alpha q_0 + \beta$ be the division of q by q_0, with $0 \le \beta < q_0$. Write

$$pH + qj_{X,\Delta}(K_X + \Delta) =$$
$$(p - \alpha p_0 - \beta)H + \beta(H + j_{X,\Delta}(K_X + \Delta)) + \alpha(p_0 H + q_0 j_{X,\Delta}(K_X + \Delta))$$

Since

$$
\begin{aligned}
p - \alpha p_0 - \beta &> p - \alpha p_0 - q_0 \\
&\ge p - \frac{q}{q_0}p_0 - q_0 \\
&> \frac{j_{X,\Delta}q - 1}{r} - \frac{q}{q_0}p_0 - q_0 \\
&= q\left(\frac{j_{X,\Delta}}{r} - \frac{p_0}{q_0}\right) - \frac{1}{r} - q_0
\end{aligned}
$$

and $\frac{jx,\Delta}{r} - \frac{p_0}{q_0} > 0$, we see that $p - \alpha p_0 - \beta$ is positive for q large enough. It follows that $B(p,q)$ is contained in $B(p_0,q_0)$.

7.36. We proceed by contradiction, assuming that either r is irrational, or $\frac{r}{jx,\Delta} = \frac{u}{v}$, with u and v relatively prime and $v > jx,\Delta(n+1)$. This implies that $\Lambda_{\frac{1}{n+1}}$ is infinite: if $\frac{r}{jx,\Delta} = \frac{u}{v}$, this set contains any positive pair (p_0,q_0) such that $vq_0 - up_0 = 1$.

Third step: there exist arbitrarily large elements (p,q) of Λ_1 such that $|pH + qjx,\Delta(K_X + \Delta)|$ is nonempty. In other words, $B_\infty \neq X$.

Let $(p,q) \in \Lambda_1$. Since $pH + (qjx,\Delta - 1)(K_X + \Delta)$ is nef and big, the logarithmic Kawamata–Viehweg vanishing theorem 7.26 implies

$$H^i(X, pH + qjx,\Delta(K_X + \Delta)) = 0 \qquad \text{for all} \quad i > 0$$

It is therefore enough to show that the polynomial

$$P(x,y) = \chi(X, xH + yjx,\Delta(K_X + \Delta))$$

does not vanish at some point of Λ_1.

Lemma 7.37 *Assume $\Lambda_{\frac{\eta}{n+1}}$ is infinite. Any polynomial $P(x,y)$ of degree at most n that vanishes on all sufficiently large elements of Λ_η is identically zero.*

PROOF. Let (p,q) be a sufficiently large element of $\Lambda_{\frac{\eta}{n+1}}$. For $1 \leq j \leq n+1$, we have $(jp, jq) \in \Lambda_\eta$, hence $P(jp, jq) = 0$. The polynomial P therefore vanishes on the line $xq = yp$. Since there are infinitely many such lines, $P = 0$. □

Since $\Lambda_{\frac{1}{n+1}}$ is infinite (7.36), it remains to show that P is nonzero to finish the proof of the third step. Since $P(x,0) = \chi(X, xH)$, this follows from Proposition 1.31.

Fix an element (p_0, q_0) of Λ_1 such that $B(p,q) = B_\infty \neq X$ for all $(p,q) \in \Lambda_1$ with $q \geq q_0$. There exists as in the proof of the base-point-free theorem 7.32 a desingularization $\pi : Y \to X$ such that

(a) the linear system $|\pi^*(p_0H + q_0jx,\Delta(K_X + \Delta)) - \sum_i r_i F_i|$ is base-point-free, where $r_i \geq 0$;

(b) the divisor $\pi^*(p_0H + (q_0jx,\Delta - 1)(K_X + \Delta)) - \sum_i p_i F_i$ is ample, where $p_i \in (0,1)$.

We write as usual

$$K_Y \sim \pi^*(K_X + \Delta) + \sum_i a_i F_i$$

with $a_i > -1$ because the pair (X, Δ) is log terminal.

The divisor $\sum_i r_i F_i$ maps onto the nonempty set B_∞, hence the r_i are not all zero. As in the proofs of Theorems 7.30 and 7.32, we may define a positive rational by

$$c = \min_i \frac{1 + a_i - p_i}{r_i}$$

and assume that the minimum is obtained for only one i. Letting F be the (smooth) corresponding F_i, we have

$$\sum_i (-cr_i + a_i - p_i)F_i = B - F$$

where $\lceil B \rceil$ is an effective divisor in which F does not appear, and which is π-exceptional because $-cr_i + a_i - p_i$ can only be positive if a_i is.

Set

$$Q(x,y) = \chi(F, \pi^*(xH + yj_{X,\Delta}(K_X + \Delta)) + \lceil B \rceil)$$

Fourth step: if

$$(p,q) \in \Lambda_{\min(1,(c+1)(aq_0 - rp_0))}$$

and $q > (c+1)q_0$, we have $Q(p,q) = 0$. However, Q is not identically zero.
For any positive integers p and q, set

$$
\begin{aligned}
N_{p,q} &= \pi^*(pH + qj_{X,\Delta}(K_X + \Delta)) - K_Y + \sum_i (-cr_i + a_i - p_i)F_i \\
&\sim \pi^*\big((p - (c+1)p_0)H + (q - (c+1)q_0)j_{X,\Delta}(K_X + \Delta)\big) \\
&\quad + (c+1)\pi^*(p_0 H + q_0 j_{X,\Delta}(K_X + \Delta)) \\
&\quad - (K_X + \Delta) + \sum_i (-cr_i - p_i)F_i \\
&\sim \pi^*\big((p - (c+1)p_0)H + (q - (c+1)q_0)j_{X,\Delta}(K_X + \Delta)\big) \\
&\quad + c\Big(\pi^*(p_0 H + q_0 j_{X,\Delta}(K_X + \Delta)) - \sum_i r_i F_i\Big) \\
&\quad + \pi^*(p_0 H + (q_0 j_{X,\Delta} - 1)(K_X + \Delta)) - \sum_i p_i F_i
\end{aligned}
$$

Assume $j_{X,\Delta}q - rp < (c+1)(j_{X,\Delta}q_0 - rp_0)$ and $q > (c+1)q_0$. We get

$$0 < j_{X,\Delta}(q - (c+1)q_0) < r(p - (c+1)p_0)$$

hence the first term of this last sum is nef, the second term is base-point-free, and the last term is ample, hence $N_{p,q}$ is ample (1.22). The vanishing theorem 7.21 implies $H^1(Y, K_Y + \lceil N_{p,q} \rceil) = 0$, hence the restriction

$$H^0(Y, \pi^*(pH + qj_{X,\Delta}(K_X + \Delta)) + \lceil B \rceil) \to$$

$$H^0(F, \pi^*(pH + qj_{X,\Delta}(K_X + \Delta)) + \lceil B \rceil) \quad (7.8)$$

is surjective. On the other hand, since

$$(N_{p,q})|_F \sim (\pi^*(pH + qj_{X,\Delta}(K_X + \Delta)) + B)|_F - K_F$$

is ample and $\lceil B \rceil|_F$ is effective with simple normal crossings, the vanishing theorem 7.21 gives

$$H^i(F, \pi^*(pH + qj_{X,\Delta}(K_X + \Delta)) + \lceil B \rceil) = 0$$

for $i > 0$, hence

$$Q(p,q) = h^0(F, \pi^*(pH + qj_{X,\Delta}(K_X + \Delta)) + \lceil B \rceil)$$

If now $0 < j_{X,\Delta}q - rp$, so that

$$(p,q) \in \Lambda_{\min(1,(c+1)(j_{X,\Delta}q_0 - rp_0))}$$

the base-locus of the left-hand side of (7.8) is $\pi^{-1}(B(p,q))$ by Lemma 7.11, which is equal to $\pi^{-1}(B_\infty)$ hence contains F. It follows that the right-hand side vanishes, i.e., $Q(p,q) = 0$.

If on the contrary $p > \frac{qj_{X,\Delta}}{r}$, the divisor $\pi^*(pH + qj_{X,\Delta}(K_X + \Delta))|_F$ is nef and $(\pi^*(pH + qj_{X,\Delta}(K_X + \Delta)) + B)|_F - K_F$ is ample, hence $Q(mp, mq) \neq 0$ by 7.31 for $m \gg 0$.

Lemma 7.37 implies

$$\Lambda_{\frac{\min(1,(c+1)(j_{X,\Delta}q_0 - rp_0))}{n+1}} = \varnothing$$

hence r is rational. When $(p,q) \in \Lambda_1$, the rational $j_{X,\Delta}q - rp$ can take at most v values, so we may choose (p_0, q_0) such that this value is maximal among large elements of Λ_1. If $(p,q) \in \Lambda_1$ and q is sufficiently large, we have therefore

$$j_{X,\Delta}q - rp \leq j_{X,\Delta}q_0 - rp_0 < (c+1)(j_{X,\Delta}q_0 - rp_0))$$

i.e.,

$$(p,q) \in \Lambda_{\min(1,(c+1)(j_{X,\Delta}q_0 - rp_0))}$$

and the fourth step implies $Q(p,q) = 0$. It follows that Q vanishes on all sufficiently large elements of Λ_1. Since $\Lambda_{\frac{1}{n+1}}$ is infinite (7.36), this contradicts Lemma 7.37.

Hypothesis 7.36 therefore leads to a contradiction in all cases. This proves the theorem. $\square$

7.9   The Cone Theorem

In Exercise 6.7.5, we saw how the cone theorem (which we had proven in Chapter 6 for smooth varieties) implies the rationality theorem 7.34. It turns out that, as M. Reid was the first to realize, the cone theorem itself is a rather easy formal consequence of the rationality theorem.

Since it is no more difficult to prove, we will state the logarithmic version of the cone theorem (we will see in Exercise 7.13.8 that this more general version is very useful even on smooth varieties).

Theorem 7.38 Let (X, Δ) be a complex[13] projective log terminal pair. The set $\mathscr{R}$ of all $(K_X + \Delta)$-negative extremal rays of $\overline{NE}(X)$ is countable and

$$\overline{NE}(X) = \overline{NE}(X)_{K_X+\Delta \geq 0} + \sum_{R \in \mathscr{R}} R$$

These rays are locally discrete in the half-space $N_1(X)_{K_X+\Delta<0}$.

In the smooth case, we proved (Theorem 6.1) that each K_X-negative extremal ray is generated by the class of a rational curve. This is still true in the singular logarithmic case, and will be proved in Corollary 7.48.

Of course, by taking $\Delta = 0$ on a projective variety with canonical singularities, we get the cone theorem in the same form as in Chapter 6.

PROOF OF THE THEOREM. We may assume that $K_X + \Delta$ is not nef. For any nef divisor M, the set $V_M = \{z \in \overline{NE}(X) \mid M \cdot z = 0\}$ is a extremal subcone of $\overline{NE}(X)$ which is nonzero if M is not ample by Theorem 1.27(b). We say that V_M is $(K_X+\Delta)$-negative if $(K_X+\Delta) \cdot z < 0$ for all $z \in V_M - \{0\}$.

First step: if M is a nef and nonample divisor such that V_M is $(K_X + \Delta)$-negative, there exists a nef divisor L such that V_L is an extremal ray contained in V_M.

Let $v = (j_{X,\Delta}(\dim(X)+1))!$ and let H be an ample divisor on X. Let m be a positive integer. Since $mM + H$ is ample by 1.22, the positive number u_m defined by

$$\frac{u_m}{v} = \sup\{t \in \mathbf{R} \mid mM + H + t(K_X + \Delta) \text{ nef}\}$$

is an integer by the rationality theorem. Since M is nef, we have $u_m \leq u_{m+1}$. Moreover, if z_0 is some nonzero element of V_M, the inequalities $(mM + H + \frac{u_m}{v}(K_X + \Delta)) \cdot z_0 \geq 0$ and $(K_X + \Delta) \cdot z_0 < 0$ imply

$$u_m \leq v \frac{(mM + H) \cdot z_0}{-(K_X + \Delta) \cdot z_0} = v \frac{H \cdot z_0}{-(K_X + \Delta) \cdot z_0}$$

It follows that the sequence (u_m) becomes stationary, equal to some integer u_∞ for $m \geq m_0$. The divisor

$$L_m = v\left(mM + H + \frac{u_\infty}{v}(K_X + \Delta)\right)$$

---

[13]Using totally different ideas, Keel has recently proved in [K] (prop. 0.6) the cone theorem for Q-factorial normal threefolds X defined over the algebraic closure of a finite field for which $|mK_X|$ is nonempty for some positive integer m.

is by construction nef and not ample for $m \geq m_0$. If $z \in V_{L_{m_0+1}}$, we have

$$0 = L_{m_0+1} \cdot z = L_{m_0} \cdot z + vM \cdot z \geq vM \cdot z \geq 0$$

hence $V_{L_{m_0+1}} \subset V_M$.

For each ample divisor H, we get a nef and nonample divisor $L_H = L_{m_0+1}$ such that $0 \neq V_{L_H} \subset V_M$. Ample classes span $N_1(X)^*_{\mathbf{R}}$, hence their restrictions span $\langle V_M \rangle^*$. If $\langle V_M \rangle$ has dimension at least 2, there exists an ample divisor H such that

$$(L_H)|_{V_M} = \left(H + \frac{u_\infty}{v}(K_X + \Delta) \right)\Big|_{V_M}$$

does not vanish, and for this divisor, V_{L_H} is strictly contained in V_M. This proves the first step by induction on the dimension of $\langle V_M \rangle$.

Second step: $\overline{NE}(X)$ *is the closure of*

$$V = \overline{NE}(X)_{K_X+\Delta\geq0} + \sum_{L\in\mathscr{L}} V_L$$

where $\mathscr{L}$ is the set of nef nonample divisors L such that V_L is a $(K_X+\Delta)$-negative extremal ray.

By Lemma 6.7(a), the boundary of the dual cone $\overline{NE}(X)^*$ meets the interior of $\overline{V}^*$ at a point that corresponds to a nonample nef $\mathbf{R}$-divisor M positive on $\overline{V}-\{0\}$. We want to show that we can find such a divisor with integral coefficients. The class of M is the limit of a sequence $([H_m])$ of classes of ample $\mathbf{Q}$-divisors. By the rationality theorem, there exists a sequence (r_m) of positive *rational* numbers such that $M_m = H_m + r_m(K_X + \Delta)$ is nef but not ample.

Since M is not ample, it vanishes at some nonzero point z_0 of $\overline{NE}(X)$. This point cannot be in V, hence $(K_X + \Delta) \cdot z_0 < 0$. This implies

$$r_m \leq \frac{H_m \cdot z_0}{-(K_X + \Delta) \cdot z_0}$$

hence the sequence (r_m) converges to 0. It follows that the sequence $([M_m])$ of classes of $\mathbf{Q}$-divisors converges to $[M]$, hence its terms are eventually in the interior of $\overline{V}^*$.

So we may assume that M has integral coefficients. Since M is not ample, V_M is nonzero. Since M is positive on $V-\{0\}$, the cone V_M is $(K_X + \Delta)$-negative. By the first step, the cone V_M contains an extremal ray V_L, and we get a contradiction since M is positive on $V_L-\{0\}$.

Third step: $\mathscr{L}$ *is countable and the $(K_X+\Delta)$-negative rays $(V_L)_{L\in\mathscr{L}}$ are locally discrete in the half-space $N_1(X)_{K_X+\Delta<0}$.*

As in the proof of the cone theorem in the smooth case (Theorem 6.1), it is enough to show that for any ample divisor H_0 and $\eta > 0$, there are finitely many such rays in the half-space $N_1(X)_{K_X+\Delta+\eta H_0<0}$. For any $L \in \mathscr{L}$, let z_L be the unique point on V_L such that $-(K_X + \Delta) \cdot z_L = 1$.

Let H be an ample divisor on X. The first step yields a nef divisor L_H with $V_L = V_{L_H}$, and $vH \cdot z_L$ is a positive integer (this is u_∞).

In the half-space $N_1(X)_{K_X+\Delta+\eta H_0<0}$, one has $H_0 \cdot z_L \leq 1/\eta$. If there are infinitely rays V_L, the z_L must accumulate by Theorem 1.27(b). This is impossible since $H \cdot z_L$ can only take discrete values for all H.

Fourth step: for any subset $\mathscr{L}'$ of $\mathscr{L}$, the cone

$$\overline{NE}(X)_{K_X+\Delta\geq0} + \sum_{L\in\mathscr{L}'} V_L$$

is closed.

This is proved exactly as the third step of the proof of Theorem 6.1. $\square$

7.10   The Contraction Theorem

The fact that extremal rays can be contracted is essential to the realization of Mori's minimal model program, but proved unattainable by the bend-and-break methods of Chapter 6. It comes now for free as an immediate consequence of the base-point-free theorem 7.32.

Recall as usual that the theorem applies when X is a projective variety with canonical singularities and $\Delta = 0$.

Theorem 7.39 *Let (X, Δ) be a projective complex[14] log terminal pair and let R be a $(K_X + \Delta)$-negative extremal ray.*

(a) *There exists an irreducible curve C on X whose class generates R.*

(b) *The contraction $c_R : X \twoheadrightarrow Y$ of R exists.*

(c) *There is an exact sequence*

$$0 \rightarrow \text{Pic}(Y) \xrightarrow{c_R^*} \text{Pic}(X) \rightarrow \mathbf{Z}$$
$$[L] \mapsto L \cdot C$$

Remarks 7.40 (1) The same theorem holds (with the same proof) for any $(K_X + \Delta)$-negative extremal subcone V of $\overline{NE}(X)$ instead of R (in which case the Picard number of $c_V(X)$ is $\rho_X - \dim(\langle V \rangle)$).

---

[14]Using totally different ideas, Keel has recently proved in [K] (Corollary, p. 256) that extremal contractions exist on $\mathbf{Q}$-factorial normal threefolds X defined over the algebraic closure of a finite field for which K_X is not nef and $|mK_X|$ is nonempty for some positive integer m.

(2) The complex appearing in item (c) of the theorem is not exact in general, as shown by the example of the second projection $c : E \times E \to E$, where E is a very general elliptic curve: the vector space $N_1(E \times E)_\mathbf{Q}$ has dimension 3, generated by the classes of $E \times \{0\}$, $\{0\} \times E$ and the diagonal ([K1], ex. II.4.16). In this basis, $\overline{NE}(E \times E)$ is the cone $xy + yz + zx \geq 0$ and $x + y + z \geq 0$, and c is the contraction of the extremal ray spanned by $(1, 0, 0)$. However, the complex

$$0 \to \mathbf{Q}(1,0,0) \to \underset{(x,y,z)}{N_1(E \times E)_\mathbf{Q}} \xrightarrow{c_*} \underset{y-z}{N_1(E)_\mathbf{Q}} \to 0$$

is not exact.

(3) Item (c) implies that there are dual exact sequences

$$0 \to N^1(Y)_\mathbf{Q} \xrightarrow{c_R^*} N^1(X)_\mathbf{Q} \xrightarrow{\text{rest}} \langle R \rangle^* \to 0$$

and

$$0 \to \langle R \rangle \to N_1(X)_\mathbf{Q} \xrightarrow{(c_R)_*} N_1(Y)_\mathbf{Q} \to 0$$

(4) We will show later (Theorem 7.46) that each $(K_X + \Delta)$-negative extremal ray R is generated by the class of a rational curve. Moreover, the exceptional locus of c_R is covered by rational curves.

PROOF OF THE THEOREM. We follow the proof of Corollary 6.9: by the fourth step of the proof of Theorem 7.38, the subcone

$$V = \overline{NE}(X)_{K_X + \Delta \geq 0} + \sum_{R' \in \mathscr{R} - \{R\}} R'$$

of $\overline{NE}(X)$ is closed, hence there exists a nef $\mathbf{R}$-divisor M that is positive on $V - \{0\}$ but vanishes at some nonzero point of $\overline{NE}(X)$, which must be in R. By the same procedure as in the second step of the proof of Theorem 7.38, we may assume that M has integral coefficients. As in the proof of Corollary 6.9(b), $mM - (K_X + \Delta)$ is ample for all m sufficiently large. By the base-point-free theorem 7.32, any sufficiently large multiple of M defines a morphism whose Stein factorization $c : X \to Y$ is the contraction of R (actually, by Proposition 7.6, there is no need to take the Stein factorization). This proves (b).

Since the contraction is unique (Proposition 1.14), any two successive large multiples of M yield isomorphic c, hence M is linearly equivalent to the pull-back of an (ample) divisor class on Y.

The morphism c^* is injective: since $c_* \mathcal{O}_X \simeq \mathcal{O}_Y$, we have $c_*(c^* \mathcal{L}) \simeq \mathcal{L}$ for any line bundle $\mathcal{L}$ on Y by the projection formula. Let now D be a Cartier divisor on X that vanishes on R. Since M is positive on $V - \{0\}$, the divisor $mM + D$ has the same property for all m sufficiently large, hence is a supporting divisor for R. By what we have just seen, $\mathcal{O}_X(mM + D)$ and

$\mathcal{O}_X(mM)$ are therefore in $c^* \operatorname{Pic}(Y)$, hence so is $\mathcal{O}_X(D)$. It follows that we have an exact sequence

$$
\begin{array}{ccccccc}
0 & \to & \operatorname{Pic}(Y) & \xrightarrow{c^*} & \operatorname{Pic}(X) & \longrightarrow & \mathbf{R} \\
& & & & [L] & \mapsto & L \cdot z
\end{array}
$$

where z is any nonzero element of R.

This shows in particular that c is not an isomorphism; Y being normal, Zariski's main theorem implies that c contracts at least one irreducible curve C, which must satisfy $M \cdot C = 0$. It follows that $[C]$ generates R and this proves (a) and (c). $\qquad\qquad\square$

7.41. Let $\pi : X \to Y$ a morphism of schemes. Relativizing the definition of Section 1.17, we say that a Cartier divisor D on X is π-*ample* if, for every coherent sheaf $\mathcal{F}$ on X, the canonical map $\pi^*(\pi_*(\mathcal{F} \otimes \mathcal{O}_X(mD))) \to \mathcal{F} \otimes \mathcal{O}_X(mD)$ is surjective for all m sufficiently large.

Equivalently, some multiple of D is induced by restriction from a closed embedding $X \subset \mathbf{P}^N \times Y$ ([Gr1], prop. 4.6.11), or Y is covered by affine subsets U such that $D|_{\pi^{-1}(U)}$ is ample.

The restriction of a π-ample divisor to every fiber of π is ample ([Gr1], prop. 4.6.13(iii)) (the converse is true if π is proper by [Gr2], th. 4.7.1).

When π is projective, there is a relative version of Kleiman's criterion (Theorem 1.27(a)): D is π-*ample if and only if it is positive on* $\overline{\mathrm{NE}}(\pi) - \{0\}$, where $\mathrm{NE}(\pi)$ is the subcone of $\mathrm{NE}(X)$ generated by classes of curves contracted by π (see 1.12).

7.42. In the case of the contraction c_R of a K_X-negative extremal ray R on a projective variety X with canonical singularities, the relative Kleiman criterion and Theorem 7.39(c) imply that $-K_X$ is c_R-ample. In particular, a general fiber is a **Q**-Fano variety (this means that it has canonical singularities (this holds by Remark 7.17(2)), and that some multiple of the canonical divisor is ample).

As discussed in Section 6.5, there are three cases:

- $\dim(c_R(X)) < \dim(X)$, the morphism c_R is a fiber contraction, and X is uniruled (see Remark 7.40(4));

- the exceptional locus of c_R is a divisor,[15] and c_R is a divisorial contraction;

- the exceptional locus of c_R has codimension at least 2, and c_R is a small contraction.[16]

---

[15]If X is **Q**-factorial, the proof of Proposition 6.10(b) shows that this divisor is irreducible. Moreover $\operatorname{Exc}(c_R) \cdot z < 0$ for any $z \in R - \{0\}$.

[16]Note, however, that the inequality of Proposition 6.10(c) no longer holds.

We already noticed in Section 6.5 that when c_R is a *small contraction*, its image is not **Q**-factorial, hence is unsuitable for the continuation of Mori's program. Instead, we would like to perform the operation described in the following definition.

Definition 7.43 *Let* $c : X \twoheadrightarrow Y$ *be a small contraction between normal projective varieties. Assume that* K_X *is* **Q**-*Cartier and* $-K_X$ *is c-ample. A* flip *of c is a small contraction* $c^+ : X^+ \to Y$ *such that*

- X^+ *is a projective normal variety;*

- K_{X^+} *is* **Q**-*Cartier and* c^+-*ample.*

The main problem here is the *existence* of a flip of a small contraction of a negative extremal ray, which is only known in dimension 3 (see [K8]).

We now examine the effect on singularities of the operations in Mori's minimal model program. For once, we revert to the standard (nonlogarithmic) case.

Proposition 7.44 *Let* X *be a locally* **Q**-*factorial complex projective variety with canonical (resp. terminal) singularities and let* $c : X \twoheadrightarrow Y$ *be the contraction of a* K_X-*negative extremal ray* R.

(a) *If c is a fiber contraction,* Y *is locally* **Q**-*factorial with Picard number* $\rho_X - 1$.[17]

(b) *If c is a divisorial contraction,* Y *is locally* **Q**-*factorial with canonical (resp. terminal) singularities, and its Picard number is* $\rho_X - 1$.

(c) *If c is a small contraction with flip* $X^+ \twoheadrightarrow Y$, *the variety* X^+ *is locally* **Q**-*factorial with canonical (resp. terminal) singularities, and its Picard number is* ρ_X.

PROOF. Let C be an irreducible curve whose class generates R (Theorem 7.39(a)). Let D be a prime Weil divisor on Y, let c^0 be the restriction of c to $X^0 = c^{-1}(Y_{\mathrm{reg}})$, and let D_X be the closure in X of $(c^0)^*(D \cap Y_{\mathrm{reg}})$.

If c is a fiber contraction, the divisor D_X is disjoint from a general fiber of c, hence has intersection 0 with C. Since X is **Q**-factorial, some multiple mD_X is a Cartier divisor, hence there exists by Theorem 7.39(c) a Cartier divisor D_Y on Y such that $mD_X \equiv c^*D_Y$. By the projection formula, mD and D_Y are linearly equivalent on Y_{reg}, hence on Y ([H1], II, prop. 6.5(b)). This proves that Y is locally **Q**-factorial, hence (a).

If c is a divisorial contraction with locus E, we have $E \cdot C < 0$ (see footnote 15, p. 207). It follows that there exists a *rational* number r such that $D_X + rE$ has intersection 0 with C. Let m be an integer such that mD

---

[17]In this case, Y might not have canonical singularities anymore.

and mrE are Cartier divisors. By Theorem 7.39(c), there exists a Cartier divisor D_Y on Y such that $m(D_X + rE) \equiv c^* D_Y$, and again, this implies (use Lemma 7.11)

$$
\begin{aligned}
\mathcal{O}_{Y_{\text{reg}}}(D_Y) &\simeq c_*^0 \mathcal{O}_{X^0}(mD_X + mrE) \\
&\simeq \mathcal{O}_{Y_{\text{reg}}}(mD) \otimes c_*^0 \mathcal{O}_{X^0}(mrE) \\
&\simeq \mathcal{O}_{Y_{\text{reg}}}(mD)
\end{aligned}
$$

hence mD and D_Y are linearly equivalent on Y. It follows that Y is locally $\mathbf{Q}$-factorial.

Let $\pi : Z \to X$ be a desingularization. Write

$$
K_Z \sim \pi^* K_X + \sum a_i E_i \sim \pi^* c^* K_Y + \sum b_i E_i
$$

where the E_i are all the exceptional divisors of $c \circ \pi$. Rewrite this equivalence as

$$
\pi^*(-(K_X - c^* K_Y)) \sim \sum (a_i - b_i) E_i
$$

Since $-K_X$ is c-ample, $-(K_X - c^* K_Y)$ is c-nef, hence $-\sum(b_i - a_i)E_i$ is $(c \circ \pi)$-nef and $(c \circ \pi)$-exceptional. Lemma 7.19 implies that $\sum(b_i - a_i)E_i$ is effective, i.e., $b_i \geq a_i$ for all i.

When X has canonical singularities, the a_i are nonnegative, hence so are the b_i, and Y has canonical singularities.

When X has terminal singularities, the a_i corresponding to π-exceptional divisors are positive, hence so are the corresponding b_i. Let E_{i_0} be a divisor on Z that is exceptional for $c \circ \pi$ but not for π. The divisor $\pi(E_{i_0})$ must then be the (irreducible) locus E of R. To show that Y has terminal singularities, we must show that if $K_X \sim c^* K_Y + aE$, where $a \geq 0$ by what we just saw, we have in fact $a > 0$. But this follows from the fact that $-K_X$ is c-ample, and (b) is proved.

Assume finally that c is a small contraction and that a flip $c^+ : X^+ \to Y$ exists. The composition $\varphi = c \circ (c^+)^{-1} : X^+ \dashrightarrow X$ is an isomorphism in codimension 1, hence induces an isomorphism of the Weil divisor class groups ([H1], II, prop. 6.5(b)). Let D^+ be a Weil divisor on X^+ and let D be the corresponding Weil divisor on X. Let r be a rational number such that $(D + rK_X) \cdot C = 0$ and let m be an integer such that mD and mrK_X are Cartier divisors. As above, there exists a Cartier divisor D_Y on Y such that $m(D + rK_X) \equiv c^* D_Y$, and

$$
mD^+ \equiv \varphi^*(mD) \equiv (c^+)^* D_Y - \varphi^*(mrK_X) \equiv (c^+)^* D_Y - mrK_{X^+}
$$

is a Cartier divisor. This proves that X^+ is locally $\mathbf{Q}$-factorial. Moreover, φ^* induces an isomorphism between $N^1(X)_{\mathbf{Q}}$ and $N^1(X^+)_{\mathbf{Q}}$, hence the Picard numbers are the same.

Let Z be a common desingularization of X and X^+, with morphisms $\pi : Z \to X$ and $\pi^+ : Z \to X^+$. Write

$$K_Z \sim \pi^* K_X + \sum a_i E_i$$
$$\sim (\pi^+)^* K_{X+} + \sum a_i^+ E_i$$

where the E_i are all the exceptional divisors of $c \circ \pi = c^+ \circ \pi^+$ (which are also all the π^+- or π-exceptional divisors). Rewrite this equivalence as

$$(\pi^+)^* K_{X+} - \pi^* K_X \sim -\sum (a_i^+ - a_i) E_i$$

Since $-K_X$ is c-ample and K_{X+} is c^+-ample, the left-hand side is $(c \circ \pi)$-nef. Since the right-hand side is $(c \circ \pi)$-exceptional, Lemma 7.19 implies that $\sum (a_i^+ - a_i) E_i$ is effective, i.e., $a_i^+ \geq a_i$ for all i. Since π^+- and π-exceptional divisors are the same, this implies (c). $\qquad \square$

The proposition shows that in the course of Mori's program (for a non-uniruled variety X), the number of divisorial contractions is bounded by $\rho_X - 1$. It does not give any information about the number of flips.

7.11   Length of Extremal Rays

We prove in this section a result of Kawamata ([Ka2]) which implies that a K_X-negative extremal ray R in the effective cone of a projective variety X with canonical singularities is generated by the class of a rational curve, or equivalently that the exceptional locus of the associated contraction contains a rational curve. Furthermore, we get an estimate on the length (see 6.11)

$$\ell(R) = \min\{-K_X \cdot \Gamma \mid \Gamma \text{ rational curve on } X \text{ with class in } R\}$$

of R, slightly weaker than in the smooth case (compare with (6.6)).

7.45. The idea of the proof is to apply the bend-and-break theorem 3.6, but we run into trouble because of the singularities. In case c_R is a fiber contraction, the proof is however very simple. A general fiber F is irreducible and nonsingular in codimension 1 (in fact, normal) and $-K_F = -K_X|_F$ is ample by 7.42 (so that F is a *singular* Fano variety in the sense of Section 5.10). Applying Theorem 3.6 to any curve contained in F_{reg} with $H = K_F$, we get that F is covered by rational curves of $(-K_F)$-degree at most $2 \dim(F)$ (compare with 6.11) hence

$$\ell(R) \leq 2 \dim(F)) = 2(\dim(X) - \dim(c_R(X)))$$

Kawamata's result generalizes this inequality to the case where the contraction c_R is birational.

Theorem 7.46 *Let X and Y be complex normal varieties and let $c : X \to Y$ be a projective morphism. Assume that Δ is an effective divisor on X such that (X, Δ) is a log terminal pair and that $-(K_X + \Delta)$ is c-ample. Every irreducible component E of $\mathrm{Exc}(c)$ is covered by rational curves Γ contracted by c and such that*

$$0 < -(K_X + \Delta) \cdot \Gamma \leq 2(\dim(E) - \dim(c(E)))$$

PROOF. If E is normal, we can proceed as in 7.45. Singularities of E will be our main concern.

Since Y is normal, $c(E)$ has codimension at least 2 in Y and $E = c^{-1}(c(E))$ (1.40). Reasoning as in the proof of Proposition 1.45, we may replace Y with the intersection of $\mathrm{codim}(c(E))$ general hyperplane sections and assume that $c(E)$ is a point: the normality of Y is preserved, so is the c-ampleness of $-K_X$, and one checks that the pair (X, Δ) remains log terminal (see [KM], Lemma 5.17; when X has canonical singularities and $\Delta = 0$, this was mentioned in Remark 7.17(2)).

Let now H be a very ample divisor on X, let e be the dimension of E, and let $\nu : \tilde{E} \to E$ be its normalization. The intersection C in $\tilde{E}$ of $e - 1$ general elements of $|\nu^* H|$ is a smooth curve contained in the smooth locus of $\tilde{E}$.

Lemma 7.47 *In the above situation, we have*

$$(K_X + \Delta) \cdot H^{e-1} \cdot E \geq K_{\tilde{E}} \cdot (\nu^* H)^{e-1} \tag{7.9}$$

In terms of the curve C, the lemma reads

$$\nu^*(K_X + \Delta) \cdot C = (K_X + \Delta)|_E \cdot \nu_* C \geq K_{\tilde{E}} \cdot C$$

The proof shows that the inequality is strict if $E \neq X$.

PROOF OF THE LEMMA. If $E = X$, there is equality in (7.9). We may therefore assume that c is birational.

Assume $e \geq 2$. If we take a general (normal) member X_H of $|H|$ and let Y_H be the normalization of $c(X_H)$, the hypersurface $E_H = E \cap X_H$ of X_H is a component of dimension $e - 1$ of the exceptional locus of $c_H : X_H \to Y_H$ and $\tilde{E}_H = \nu^{-1}(H)$ is normal, hence is the normalization of E_H. Set $\Delta_H = \Delta|_H$; since

$$
\begin{aligned}
(K_{X_H} + \Delta_H) \cdot H^{e-2} \cdot E_H &= (K_X + \Delta + H)|_H \cdot H^{e-2} \cdot E|_H \\
&= (K_X + \Delta) \cdot H^{e-1} \cdot E + H^e \cdot E
\end{aligned}
$$

and

$$
\begin{aligned}
K_{\tilde{E}_H} \cdot (\nu^* H)^{e-2} &= (K_{\tilde{E}} + \nu^* H)|_{\tilde{E}_H} \cdot (\nu^* H)^{e-2} \\
&= K_{\tilde{E}} \cdot (\nu^* H)^{e-1} + H^e \cdot E
\end{aligned}
$$

it is the same to prove the lemma for c or for c_H. We may therefore assume $e = 1$ (but we lose the c-ampleness of $-(K_X + \Delta)$, of course) and prove

$$(K_X + \Delta) \cdot E > \deg(K_{\tilde{E}})$$

Assume to the contrary

$$(K_X + \Delta) \cdot E \le \deg(K_{\tilde{E}})$$

(note that the left-hand side is a rational number, whereas the right-hand side is an integer). Let D_E be a divisor on E_{reg} such that $\nu^* D_E \equiv K_{\tilde{E}}$. There is an injective trace map[18] $\nu_* \omega_{\tilde{E}} \to \omega_E$. The projection formula implies

$$h^0(E, \omega_E(-D_E)) \ge h^0(E, \nu_* \omega_{\tilde{E}}(-D_E)) = h^0(\tilde{E}, \omega_{\tilde{E}}(-\nu^* D_E)) = 1$$

hence

$$H^1(E, D_E) \ne 0 \tag{7.10}$$

by duality. Note for future reference that

$$D_E - (K_X + \Delta)|_E \quad \text{has nonnegative degree.} \tag{7.11}$$

After shrinking Y, we may assume that it is contractible and Stein[19] and that c induces an isomorphism over the complement of the point $0 = c(E)$. The curve E is a component of the fiber $X_0 = c^{-1}(0)$. The sheaves $R^i c_* \mathcal{O}_X$ and $R^i c_* \mathbf{Z}$, for $i > 0$, are supported at 0 and the corresponding Leray exact sequences yield isomorphisms

$$H^i(X, \mathbf{Z}) \simeq H^0(Y, R^i c_* \mathbf{Z}) \simeq H^i(X_0, \mathbf{Z})$$

and

$$H^i(X, \mathcal{F}) \simeq H^0(Y, R^i c_* \mathcal{F}) \tag{7.12}$$

for any coherent sheaf $\mathcal{F}$ on X and integer i.

We will make the simplifying assumption[20] that X_0 has dimension 1. Using (7.12), this implies $H^2(X, \mathcal{F}) = 0$ for any coherent sheaf $\mathcal{F}$ on

---

[18]If $\pi : X \to Y$ is a finite surjective morphism between projective integral schemes over a field and ω_Y is a dualizing sheaf for Y, the sheaf $\mathcal{H}om_Y(\pi_* \mathcal{O}_X, \omega_Y)$ is a sheaf of $\pi_* \mathcal{O}_X$-modules which can be written as $\pi_* \omega_X$, where ω_X is a dualizing sheaf for X. Taking the image of 1 yields a trace morphism $\pi_* \omega_X \to \omega_Y$. This morphism is injective because $\pi_* \omega_X$ is torsion free (see [H1], III, ex. 6.10 and 7.2, or [Bo], §9, for an algebraic and elementary treatment of duality, or also [KM], §5.5).

[19]So that the higher cohomology of $\mathbf{Z}$ and of any coherent sheaf vanishes.

[20]This holds, for example, when E is a top-dimensional component of the exceptional locus of c, and this is enough to prove Corollary 7.48. In general, one may use a neat

X. The exponential exact sequences for X and X_0 induce a commutative diagram

$$
\begin{array}{ccccccc}
H^1(X, \mathscr{O}_X) & \to & \mathrm{Pic}(X) & \to & H^2(X, \mathbf{Z}) & \to & H^2(X, \mathscr{O}_X) = 0 \\
\downarrow & & \downarrow & & \downarrow{\wr} & & \downarrow \\
H^1(X_0, \mathscr{O}_{X_0}) & \to & \mathrm{Pic}(X_0) & \to & H^2(X_0, \mathbf{Z}) & \to & H^2(X_0, \mathscr{O}_{X_0}) = 0 \\
\downarrow & & & & & & \\
H^2(X, I_{X_0/X}) = 0 & & & & & &
\end{array}
$$

hence a surjection $\mathrm{Pic}(X) \twoheadrightarrow \mathrm{Pic}(X_0)$. Since E meets the rest of the components of X_0 in a finite set, there is a Cartier divisor D on X which restricts to D_E on E and is as ample as we wish on $\overline{X_0 - E}$. In particular, the $\mathbf{Q}$-Cartier divisor $D - (K_X + \Delta)$ is c-nef by (7.11), and c-big because c is birational. Since X has canonical singularities, by the relative logarithmic Kawamata–Viehweg vanishing theorem (Exercise 7.13.4), the sheaves $R^i c_*(\mathscr{O}_X(D))$ vanish for $i > 0$, hence also $H^i(X, D)$ by (7.12).

Part of the long exact sequence in cohomology associated with the exact sequence

$$0 \to I_{E/X}(D) \to \mathscr{O}_X(D) \to \mathscr{O}_E(D_E) \to 0$$

reads

$$0 = H^1(X, D) \to H^1(E, D_E) \to H^2(X, I_{E/X}(D)) = 0$$

It contradicts (7.10) and proves the lemma. $\qquad\square$

We now go back to the proof of the theorem. Since the left-hand side of (7.9) is negative, Theorem 3.6 implies that $\tilde{E}$ contains a rational curve through any given point of C. In particular, E is covered by rational curves. More precisely, since $-\nu^*(K_X + \Delta)$ is ample on $\tilde{E}$, there exists through any point x of $\tilde{E}$ a rational curve Γ on $\tilde{E}$ with

$$-\nu^*(K_X + \Delta) \cdot \Gamma \le 2e \, \frac{-\nu^*(K_X + \Delta) \cdot C}{-K_{\tilde{E}} \cdot C} \le 2\epsilon$$

by the lemma. This finishes the proof of the theorem. $\qquad\square$

The following is an immediate consequence of the theorem. It applies of course to a projective variety X with canonical singularities and $\Delta = 0$.

---

trick of Kawamata's to reduce to the case where X_0 is actually irreducible. This goes roughly as follows (see [Ka5], (2.3.3)): take a very ample divisor on X that meets X_0 transversely and shrink X to an analytic neighborhood of E so that this divisor has a component that meets only the other components of X_0, not E. The holomorphic map to a projective space associated with the sections of (a multiple of) this divisor contracts only E. Its image might not be algebraic, but the vanishing theorem (Exercise 7.13.4) that we use below is still valid in this context (see [Na], 3.6) and the proof therefore proceeds in the same way.

Corollary 7.48 *Let (X, Δ) be a complex projective log terminal pair. The length of any $(K_X + \Delta)$-negative extremal ray is at most $2\dim(X)$.*

In other words, any such ray is generated by the class of a rational curve Γ such that

$$-(K_X + \Delta) \cdot \Gamma \leq 2\dim(X)$$

We end the section with the final form of the cone theorem (see, however, Theorem 7.51).

Theorem 7.49 (Logarithmic cone theorem) *Let (X, Δ) be a projective log terminal pair. There exists a countable family $(\Gamma_i)_{i \in I}$ of rational curves on X such that*

$$0 < -(K_X + \Delta) \cdot \Gamma_i \leq 2\dim(X)$$

and

$$\overline{NE}(X) = \overline{NE}(X)_{K_X + \Delta \geq 0} + \sum_{i \in I} \mathbf{R}^+ [\Gamma_i]$$

where the $\mathbf{R}^+[\Gamma_i]$ are all the (distinct) extremal rays of $\overline{NE}(X)$ that meet $N_1(X)_{K_X + \Delta < 0}$. These rays are locally discrete in that half-space.

Recall that these extremal rays can be contracted (Theorem 7.39). We give in Exercises 7.13.8 and 7.13.9 two applications of this theorem that clearly show the benefits of logarithmic geometry.

All the results in [KaMM] and in [KM] (as of §2.2) are stated in the logarithmic framework. Mori's program can be described in the same terms (see [KM], §3.7) although the concept of minimal model of a pair remains unclear.

7.12 Existence of Flips and Extensions of the Minimal Model Program

We go back for simplicity to the standard case of the minimal model program, dealing with a single projective variety with canonical singularities, although everything can be done without any major changes in the logarithmic framework (see [KM], §3.7).

Contrary to what its title suggests, we will not prove much in this section towards the existence of flips. The aim is rather to translate the problem of the existence of flips into a problem of the same type as the key problem 7.2, and to explain how this new problem can be seen as a generalization of the key problem, and of the minimal model program.

Proposition 7.50 *Let X and Y be normal projective varieties and let $c : X \to Y$ be a small contraction. Assume that K_X is $\mathbf{Q}$-Cartier and that*

$-K_X$ is c-ample. Let j be a positive integer such that jK_X is a Cartier divisor.

A flip $c^+ : X^+ \to Y$ of c exists if and only if the $\mathcal{O}_Y$-algebra

$$\mathcal{R} = \bigoplus_{m \geq 0} c_* \mathcal{O}_X(mjK_X)$$

is finitely generated. If this is the case, the Y-scheme X^+ is isomorphic to $\mathbf{Proj}(\mathcal{R})$.

PROOF. Assume that a flip $c^+ : X^+ \to Y$ of c exists. To prove that $\mathcal{R}$ is finitely generated, we may replace j with any positive multiple (see 7.5), so we may assume that jK_{X^+} is a Cartier divisor. Since K_{X^+} is c^+-ample, the graded $\mathcal{O}_Y$-algebra

$$\mathcal{R}^+ = \bigoplus_{m \geq 0} c_*^+ \mathcal{O}_{X^+}(mjK_{X^+})$$

is finitely generated and $X^+ \simeq \mathbf{Proj}(\mathcal{R}^+)$ (by item (b) of [Gr1], prop. 4.6.3, there is an open dominating immersion from X^+ to $\mathbf{Proj}(\mathcal{R})$, which is also closed since X^+ is proper over Y). Since X^+ and X are normal and $(c^+)^{-1} \circ c$ induces an isomorphism between open subsets of X and X^+ whose complements have codimension at least 2, we have $c_*^+ \mathcal{O}_{X^+}(mjK_{X^+}) \simeq c_* \mathcal{O}_X(mjK_X)$ and $\mathcal{R}^+$ is isomorphic to $\mathcal{R}$.

Assume conversely that $\mathcal{R}$ is a finitely generated $\mathcal{O}_Y$-algebra, and set $X^+ = \mathbf{Proj}(\mathcal{R})$, with canonical morphism $c^+ : X^+ \to Y$. This projective variety is constructed very much as in 7.3: if m is a positive integer such that $\mathcal{R}^{(m)}$ is generated by $\mathcal{R}_m = c_* \mathcal{O}_X(mjK_X)$, and U is any affine open subset of Y, the U-scheme $(c^+)^{-1}(U)$ is the image of the rational map

$$\varphi^U_{mjK_X} : c^{-1}(U) \dashrightarrow \mathbf{P}(H^0(c^{-1}(U), mjK_X)) \simeq \mathbf{P}(\mathcal{R}_m(U))$$

associated with the global sections of the invertible sheaf $\mathcal{O}_{c^{-1}(U)}(mjK_X)$ ([H1], II, th. 7.1). One obtains the Y-variety X^+ and the surjective rational Y-map $\varphi : X \dashrightarrow X^+$ by gluing the $(c^+)^{-1}(U)$ and the $\varphi^U_{mjK_X}$. There is also a c^+-very ample sheaf $\mathcal{O}_{X^+}(1)$ that restricts to $\mathcal{O}_{\mathbf{P}(\mathcal{R}_m(U))}(1)$ on $(c^+)^{-1}(U)$ ([H1], p. 160). It satisfies $\varphi^* \mathcal{O}_{X^+}(1) \simeq \mathcal{O}_X(mjK_X)$.

Since c is birational and φ is surjective, φ and c^+ must be birational. Exactly as in Lemma 7.10, one proves that X^+ is normal and that φ induces an isomorphism between an open subset X_0 of X, and a smooth open subset X_0^+ of X^+ whose complement has codimension at least 2. Since $\varphi^* \mathcal{O}_{X^+}(1) \simeq \mathcal{O}_X(mjK_X)$, the invertible sheaves $\mathcal{O}_{X^+}(1)$ and $\mathcal{O}_{X^+}(mjK_{X^+})$ are isomorphic on X_0^+ hence on X^+ since the complement has codimension at least 2 and X^+ is normal. It follows that K_{X^+} is c^+-ample.

It remains to show that c^+ is a small contraction: let E^+ be a divisor on X^+ and let E be the closure in X of $\varphi|_{X_0}^{-1}(E^+ \cap X_0^+)$. Since c is a small contraction, $c^+(E^+) = c(E)$ is a divisor in Y. This proves that c^+ is a small contraction. $\square$

This results suggests that, as often in algebraic geometry, one should study the relative situation, i.e., the case of a proper morphism $\pi : X \to S$. We still impose conditions on the singularities of X (not of the fibers, as one might have expected) and we are interested in the numerical properties of the canonical divisor K_X.

We have already defined π-ampleness in 7.41 and π-nefness in footnote 5, p. 181. A Cartier divisor D on X is π-big if $\mathrm{rank}(\pi_* \mathcal{O}_X(mD)) > cm^n$ for some positive constant c as m goes to infinity, where n is the dimension of a general fiber of π (i.e., $n = \dim(X) - \dim(\pi(X))$).

There is a relative Kawamata–Viehweg vanishing theorem (see Exercise 7.13.2) and a relative base-point-free theorem whose proof runs along the same lines as the proof of Theorem 7.32 (see [KaMM], th. 3.1.1, or [KM], th. 3.24): it states that if $\pi : X \to S$ is a proper morphism between complex projective varieties, if X has canonical singularities, and if D is a π-nef Cartier divisor on X such that $aD - K_X$ is π-nef and π-big for some positive rational a, the canonical map

$$\pi^*(\pi_* \mathcal{O}_X(mD)) \to \mathcal{O}_X(mD)$$

is surjective for all integers m sufficiently large (we say that mD is π-free).

There is also a relative cone theorem that describes the relative closed cone $\overline{\mathrm{NE}}(\pi)$ (defined in 1.12) and a relative contraction theorem that says that a K_X-negative extremal ray of $\overline{\mathrm{NE}}(\pi)$ can be contracted to an S-variety.

The minimal model program in this case aims at finding, through a series of well-understood steps, a relative minimal model of $\pi : X \to S$, that is a proper morphism $\pi' : X' \to S$, where X' has canonical singularities, is S-birational to X, and $K_{X'}$ is π'-nef.

If the canonical $\mathcal{O}_S$-algebra

$$\mathscr{R}(X/S, K_X) = \bigoplus_{m \geq 0} \pi_* \mathcal{O}_X(mK_X)$$

is finitely generated, one may define the *relative canonical model* of X as the S-scheme $\mathbf{Proj}(\mathscr{R}(X/S, K_X))$. The relative base-point-free theorem implies that this holds if K_X is π-nef and π-big (see [KaMM], th. 3.3.1). Note that the existence of the flip of a small contraction is a problem of this type.

Finally, as the reader will have guessed, one can combine the logarithmic and the relative situations. Everything is stated in this degree of generality in [KM] (see §3.6 and following). The cone theorem takes the following form ([KM], th. 3.25).

Theorem 7.51 (Relative logarithmic cone theorem) *Let (X, Δ) be a projective log terminal pair and let $\pi : X \to S$ be a projective morphism.*

(a) *There exists a countable family $(\Gamma_i)_{i \in I}$ of rational curves on X contracted by π such that*

$$0 < -(K_X + \Delta) \cdot \Gamma_i \leq 2 \dim(X)$$

and

$$\overline{\mathrm{NE}}(\pi) = \overline{\mathrm{NE}}(\pi)_{K_X + \Delta \geq 0} + \sum_{i \in I} \mathbf{R}^+ [\Gamma_i]$$

where the $\mathbf{R}^+[\Gamma_i]$ are all the (distinct) extremal rays of $\overline{\mathrm{NE}}(\pi)$ that meet $N_1(X)_{K_X + \Delta < 0}$. These rays are locally discrete in that half-space.

(b) *Each ray $\mathbf{R}^+[\Gamma_i]$ can be contracted by a morphism c_i to a projective S-variety Y and there is an exact sequence*

$$
\begin{array}{ccccccc}
0 & \to & \mathrm{Pic}(Y) & \xrightarrow{c_i^*} & \mathrm{Pic}(X) & \to & \mathbf{Z} \\
 & & & & [L] & \mapsto & L \cdot \Gamma_i
\end{array}
$$

7.13 Exercises

1. Let X be a projective surface with canonical singularities and let $\pi : Y \to X$ be a *minimal* desingularization. Write

$$K_Y \sim \pi^* K_X + \sum_{i=1}^{m} a_i E_i$$

where the $E_1, \ldots, E_m$ are the irreducible curves on Y contracted by π and $a_1, \ldots, a_m$ are nonnegative rational numbers.

(a) Show that $a_1, \ldots, a_m$ all vanish: if X has terminal singularities, X is smooth.

(b) Show that $E_1, \ldots, E_m$ are smooth rational curves with self-intersection -2.

2. **A relative Kawamata–Viehweg vanishing theorem.** Let X and Y be complex projective varieties, with X smooth, and let $\pi : X \to Y$ be a morphism. Let D be a π-nef and π-big $\mathbf{Q}$-divisor on X whose fractional part has simple normal crossings. We want to prove

$$R^i \pi_*(\mathcal{O}_X(K_X + \lceil D \rceil)) = 0$$

for all $i > 0$. Let $\mathcal{F}_i$ be this coherent sheaf on Y.

(a) Let H be any ample divisor on Y such that $H^j(Y, \mathscr{F}_i(H)) = 0$ for all $j > 0$ and i. Use the Leray spectral sequence associated with π to show

$$H^0(Y, \mathscr{F}_i(H)) \simeq H^i(X, K_X + \lceil D \rceil + \pi^* H)$$

(b) Use the Kawamata–Viehweg vanishing theorem 7.21 to conclude.

3. **The logarithmic Kawamata–Viehweg vanishing theorem.** We want to prove Theorem 7.26: let (X, Δ) be a projective log terminal pair (see Definition 7.24) and let D be a nef and big **Q**-divisor on X such that $K_X + \Delta + D$ is an integral Cartier divisor.

Prove

$$H^i(X, K_X + \Delta + D) = 0$$

(*Hint:* If $\pi : Y \to X$ is a desingularization, apply Theorem 7.21 and Exercise 7.13.2 to $\pi^* D$.)

4. **A relative logarithmic Kawamata–Viehweg vanishing theorem.** Let (X, Δ) be a projective log terminal pair (see Definition 7.24), let D be a π-nef and π-big **Q**-Cartier **Q**-divisor on X such that $K_X + \Delta + D$ is an integral Cartier divisor, and let $\pi : X \to Y$ be a morphism. Prove

$$R^i \pi_* (\mathscr{O}_X(K_X + \Delta + D)) = 0$$

for all $i > 0$.

As an application, assume that $-(K_X + \Delta)$ is π-ample (this is the case for the contraction of a $(K_X + \Delta)$-negative extremal ray) and that the exceptional locus E of π is one-dimensional. Prove that E is a tree of smooth rational curves.

5. **A rationality theorem for nef divisors.** Let X be a projective variety with canonical singularities and index j_X, whose canonical divisor is not nef. Let M be a nef Cartier divisor on X. Show that

- the number
$$r = \sup\{t \in \mathbf{R} \mid M + t K_X \text{ nef}\}$$
is rational;

- one can write
$$\frac{r}{j_X} = \frac{u}{v}$$
where u and v are relatively prime integers and satisfy $0 < v \le 2 j_X \dim(X)$;

- there exists a K_X-negative extremal ray R of $\overline{\mathrm{NE}}(X)$ such that

$$(M + rK_X) \cdot R = 0$$

(*Hint:* Use the cone theorem 7.38 and Corollary 7.48, and proceed as in Exercise 6.7.5.)

6. Let D be an effective divisor on a smooth projective variety X such that the multiplier ideal $\mathscr{I}_{\lambda D}$ (see §7.5) is trivial for all $\lambda \in (0,1) \cap \mathbf{Q}$. Show that the locus $\mathrm{Sing}_k(D)$ of points of X that have multiplicity at least k on D has codimension at least k in X. (*Hint:* Construct a desingularization of X by first blowing up a top-dimensional irreducible component of $\mathrm{Sing}_k(D)$.)

7. Recall that a complex principally polarized abelian variety (A, Θ) is a complex abelian variety A with an ample effective divisor Θ such that $h^0(A, \Theta) = 1$. We want to show that for any positive integer k,

$$\mathrm{codim}_A(\mathrm{Sing}_k(\Theta)) \geq k$$

   (a) Let λ be a rational number between 0 and 1 and let Z be the subscheme of A defined by the multiplier ideal $\mathscr{I}_{\lambda\Theta}$. Use Nadel's vanishing theorem to prove that the restriction

$$H^0(A, \mathscr{O}_A(\Theta)) \to H^0(Z, \mathscr{O}_Z(\Theta))$$

   is surjective.

   (b) Deduce that Z is empty. The result now follows from Exercise 7.13.6.

8. Let X be a smooth projective variety of general type that contains no rational curves. We want to show that K_X is ample and proceed by contradiction.

   (a) Show that K_X is nef.

   (b) Show that there exist on X a curve C and an effective divisor D such that

$$K_X \cdot C = 0 \quad \text{and} \quad D \cdot C < 0 \qquad (7.13)$$

   (*Hint:* Use the base-point-free theorem 7.32.)

   (c) Show that (7.13) implies the existence of a rational curve on X (*Hint:* Use Theorem 7.49.)

9. Let X be a projective variety with canonical singularities, of general type.[21]

---

[21]We say that a normal projective variety X is *of general type* if some (hence all) desingularization is. When X has canonical singularities, this is equivalent by 7.15 to K_X big.

(a) Show that there exists an effective Cartier divisor D such that $K_X - D$ is ample. (*Hint:* Use Exercise 1.12.5.)

(b) Show that the pair $(X, \varepsilon D)$ is log terminal for ε rational positive small enough.

(c) Show that there are finitely many K_X-negative extremal rays in $\overline{NE}(X)$. (*Hint:* Use Theorem 7.49.)

(d) Construct a smooth projective variety X with dimension n and Kodaira dimension $n - 2$ (this means $\dim(\varphi_{mK_X}(X)) = n - 2$ for all m large enough) for which there are infinitely many K_X-negative extremal rays in $\overline{NE}(X)$. (*Hint:* Use the construction in 6.6.)

References

[AKMW]   Abramovich, D., Karu, K., Matsuki, K., and Włodarczyk, J., Torification and factorization of birational maps, preprint.

[AJ]   Abramovich, D. and de Jong, A. J., Smoothness, semistability, and toroidal geometry, *J. Alg. Geom.* **6** (1997), 789–801.

[AK]   A. Albano, A. and Katz, S., Lines on the Fermat quintic three-fold and the infinitesimal generalized Hodge conjecture, *Trans. Amer. Math. Soc.* **324** (1991), 353–368.

[AW1]   Andreatta, M. and Wiśniewski, J., On contractions of smooth varieties, *J. Alg. Geom.* **7** (1998), 253–312.

[AW2]   Andreatta, M. and Wiśniewski, J., Contractions of smooth varieties. II. Computations and applications, *Boll. Unione Mat. Ital. Sez. B Artic. Ric. Mat. (8)* **1** (1998), 343–360.

[AS]   Angehrn, U. and Siu, Y.-T., Effective freeness and point separation, *Invent. Math.* **122** (1995), 291–308.

[Ba]   Batyrev, V., Boundedness of the degree of multidimensional toric Fano varieties (in Russian with English abstract), *Vestnik Moskov. Univ. Ser. I Mat. Mekh.* (1982) 22–27, 76–77; English transl.: *Moscow Univ. Math. Bull.* **37** (1982), 28–33.

[B1]   Beauville, A., Some remarks on Kähler manifolds with $c_1 = 0$, *Classification of Algebraic and Analytic Manifolds (Katata, 1982)*, Progr. Math. **39**, Birkhäuser, Boston, 1983.

[B2]   Beauville, A., Sur les hypersurfaces dont les sections hyperplanes sont à module constant, *The Grothendieck Festschrift, Vol. I*, Progr. Math. **86**, Birkhäuser, Boston, 1990.

[Be]   Besse, A., *Einstein Manifolds*, Ergebnisse der Mathematik und ihrer Grenzgebiete **10**, Springer-Verlag, New York, 1987.

[BC]   Bishop, R. and Crittenden, R., *Geometry of Manifolds*, Pure and Applied Mathematics **XV**, Academic Press, New York, 1964.

[BP]   Bogomolov, F. and Pantev, T., Weak Hironaka theorem, *Math. Res. Lett.* **3** (1996), 299–307.

[Bo]   Bourbaki, N., *Algèbre commutative*, Éléments de Mathématique, Chapitre 10, Masson, Paris, 1998.

[Bou1]   Bourguignon, J.-P., Premières formes de Chern des variétés kählériennes compactes, d'après E. Calabi, T. Aubin et S. T. Yau, *Séminaire Bourbaki, 1977/78*, Exp. 507, Lectures Notes in Mathematics **710**, Springer-Verlag, New York, 1979.

[Bou2]   Bourguignon, J.-P., Métriques d'Einstein–Kähler sur les variétés de Fano : obstructions et existence (d'après Y. Matsushima, A. Futaki, S. T. Yau, A. Nadel et G. Tian), *Séminaire Bourbaki, 1996/97*, Exp. 830, Astérisque **245** (1997), 277–305.

[Br]   Bredon, A., *Sheaf Theory*, Springer-Verlag, New York, 1997.

[C1]   Campana, F., Coréduction algébrique d'un espace analytique compact faiblement kählérien, *Invent. Math.* **63** (1981), 187–223.

[C2]   Campana, F., On the twistor spaces of class $\mathscr{C}$, *J. Diff. Geom.* **33** (1991), 541–549.

[C3]   Campana, F., Connexité rationnelle des variétés de Fano, *Ann. Sci. École Norm. Sup.* **25** (1992), 539–545.

[C4]   Campana, F., Remarques sur le revêtement universel des variétés kählériennes compactes, *Bull. Soc. Math. Fr.* **122** (1994), 255–284.

[C5]   Campana, F., *G*-connectedness of compact Kähler manifolds, I, *Algebraic Geometry: Hirzebruch 70, Warsaw, 1998*, P. Pragacz, M. Szurek, J. Wiśniewski, eds., Contemp. Math. **241**, Amer. Math. Soc., Providence, 1999.

[Ca]   Catanese, F., Chow varieties, Hilbert schemes and moduli spaces of surfaces of general type, *J. Alg. Geom.* **1** (1992), 561–596.

[CC]   Ceresa, G. and Collino, A., Some remarks on algebraic equivalence of cycles, *Pacific J. Math.* **105** (1983), 285–290.

[Ch]   Chen, X., Unirationality of Fano Varieties, *Duke Math. J.* **90** (1997), 63–71

[CKM]   Clemens, H., Kollár, J., and Mori, S., *Higher-Dimensional Complex Geometry*, Astérisque **166** (1988).

[D]   Debarre, O., Variétés de Fano, *Séminaire Bourbaki 1996/97*, Exp. 827, Astérisque **245** (1997), 197–221.

[DM]   Debarre, O. and Manivel, L., Sur la variété des espaces linéaires contenus dans une intersection complète, *Math. Ann.* **312** (1998), 549–574.

[De1]   Demailly, J.-P., A numerical criterion for very ample line bundles, *J. Diff. Geom.* **37** (1993), 323–374.

[De2]   Demailly, J.-P., Singular Hermitian metrics on positive line bundles, *Complex Algebraic Varieties, Proceedings, 1990*, Lectures Notes in Mathematics **1507**, Springer-Verlag, New York, 1992.

[DPS]   Demailly, J.-P., Peternell, T., and Schneider, M., Compact complex manifolds with numerically effective tangent bundles, *J. Alg. Geom.* **3** (1994), 295–345.

[Dr]   Druel, S., Structures de contact sur les variétés algébriques de dimension 5, *C. R. Acad. Sci.* **327** (1998), 365–368.

[EL]   Ein, L. and Lazarsfeld, R., Singularities of theta divisors, and the birational geometry of irregular varieties, *J. Amer. Math. Soc.* **10** (1996), 243–258.

[Ek]   Ekedahl, T., Canonical models of surfaces of general type in positive characteristic, *Inst. Hautes Études Sci. Publ. Math.* **67** (1988), 97–144.

[El]   Elkik, R., Rationalité des singularités canoniques, *Invent. Math.* **67** (1988), 97–144.

[EV]   Esnault, H. and Viehweg, E., Logarithmic de Rham complexes and vanishing theorems, *Invent. Math.* **86** (1986), 161–194.

[F]   Flenner, H., Rational singularities, *Arch. Math.* **36** (1981), 35–44.

[Fr]   Francia, P., Some remarks on minimal models. I, *Comp. Math.* **40** (1980), 301–313.

[G]   Griffiths, P., On the periods of certain rational integrals, II, *Ann. of Math.* **90** (1969), 496–541.

[Gr1]   Grothendieck, A., Éléments de Géométrie Algébrique II, *Inst. Hautes Études Sci. Publ. Math.* **8**, 1965.

[Gr2]   Grothendieck, A., Éléments de Géométrie Algébrique III, 1, *Inst. Hautes Études Sci. Publ. Math.* **11**, 1966.

[Gr3]   Grothendieck, A., Éléments de Géométrie Algébrique IV, 2, *Inst. Hautes Études Sci. Publ. Math.* **24**, 1965.

[Gr4]   Grothendieck, A., Éléments de Géométrie Algébrique IV, 3, *Inst. Hautes Études Sci. Publ. Math.* **28**, 1966.

[Gr5]   Grothendieck, A., Techniques de construction et théorèmes d'existence en géométrie algébrique IV : les schémas de Hilbert, *Séminaire Bourbaki, 1960/61*, Exp. 221, Astérisque hors série **6**, Soc. Math. Fr., 1997.

[Gr6]   Grothendieck, A., *Cohomologie locale des faisceaux cohérents et théorèmes de Lefschetz locaux et globaux (SGA 2)*, North-Holland, Amsterdam, 1968.

[HMP]   Harris, J., Mazur, B., and Pandharipande, R., Hypersurfaces of low degree, *Duke Math. J.* **95** (1998), 125–160.

[H1]   Hartshorne, R., *Algebraic Geometry*, Graduate Texts in Mathematics **52**, Springer-Verlag, New York, 1977.

[H2]   Hartshorne, R., *Ample Subvarieties of Algebraic Varieties*, Lecture Notes in Mathematics **156**, Springer-Verlag, New York, 1970.

[H3]   Hartshorne, R., *Residues and Duality*, Lecture Notes in Mathematics **20**, Springer-Verlag, New York, 1966.

[Hi1]   Hironaka, H., Resolution of singularities of an algebraic variety over a field of characteristic zero. I, II, *Ann. of Math.* **79** (1964), 109–326.

[Hi2]   Hironaka, H., On resolution of singularities (characteristic zero), *Proc. Internat. Congr. Mathematicians (Stockholm, 1962)*, Inst. Mittag-Leffler, Djursholm, 1963.

[I1]   Iskovskikh, V., Fano 3–folds I (in Russian), *Izv. Akad. Nauk SSSR Ser. Mat.* **41** (1977), 516–562; English transl.: *Math. USSR Izvestiya* **11** (1977), 485–527.

[I2]   Iskovskikh, V., Fano 3-folds II (in Russian), *Izv. Akad. Nauk SSSR Ser. Mat.* **42** (1978), 506–549; English transl.: *Math. USSR Izvestiya* **12** (1978), 469–506.

[Ka1]   Kawamata, Y., Minimal models and the Kodaira dimension of algebraic fiber spaces, *J. für die reine und angewandte Math.* **363** (1985), 1–46.

[Ka2]   Kawamata, Y., On the length of an extremal rational curve, *Invent. Math.* **105** (1991), 609–611.

[Ka3]   Kawamata, Y., A generalization of Kodaira–Ramanujam's vanishing theorem, *Math. Ann.* **261** (1982), 43–46.

[Ka4]   Kawamata, Y., Boundedness of **Q**-Fano threefolds, *Proceedings of the International Conference on Algebra, (Novosibirsk, 1989)*, Contemp. Math. **131**, Part 3, Amer. Math. Soc., Providence, 1992.

[Ka5]   Kawamata, Y., Small contractions of four dimensional algebraic manifolds, *Math. Ann.* **284** (1989), 595–600.

[KaMM]   Kawamata, Y., K. Matsuda, K., and Matsuki, K., Introduction to the minimal model problem, *Algebraic Geometry, Sendai, 1985*, Adv. Stud. Pure Math. **10**, North-Holland, Amsterdam, 1987.

[K]   Keel, S., Basepoint freeness for nef and big line bundles in positive characteristic, *Ann. of Math.* **149** (1999), 253–286.

[Kle]   Kleiman, S., Towards a numerical theory of ampleness, *Ann. of Math.* **84** (1966), 293–344.

[K1]   Kollár, J., *Rational Curves on Algebraic Varieties*, Ergebnisse der Mathematik und ihrer Grenzgebiete **32**, Springer-Verlag, New York, 1996.

[K2]   Kollár, J., The structure of algebraic threefolds: an introduction to Mori's program, *Bull. Amer. Math. Soc.* **17** (1987), 211–273.

[K3]   Kollár, J., Effective base point freeness, *Math. Ann.* **296** (1993), 595–605.

[K4]   Kollár, J., *Shafarevich Maps and Automorphic Forms*, M. B. Porter Lectures, Princeton University Press, Princeton, 1995.

[K5]   Kollár, J., Vanishing theorems for cohomology groups, *Algebraic Geometry, Bowdoin 1985*, Proc. Symp. Pure Math. **46**, Amer. Math. Soc., Providence, 1987.

[K6]   Kollár, J., Flops, *Nagoya Math. J.* **113** (1989), 15–36.

[K7]   Kollár, J., Singularities of pairs, *Algebraic Geometry—Santa Cruz 1995*, Proc. Symp. Pure Math. **62**, Part 1, Amer. Math. Soc., Providence, 1997.

[K8]   Kollár, J., ed., *Flips and Abundance for Algebraic Threefolds*, Astérisque **211** (1993).

[KMM1] Kollár, J., Miyaoka, Y., and Mori, S., Rational curves on Fano varieties, *Proc. Alg. Geom. Conf. Trento*, Lecture Notes in Mathematics **1515**, Springer-Verlag, New York, 1992.

[KMM2] Kollár, J., Miyaoka, Y., and Mori, S., Rationally connected varieties, *J. Alg. Geom.* **1** (1992), 429–448.

[KMM3] Kollár, J., Miyaoka, Y., and Mori, S., Rational connectedness and boundedness of Fano manifolds, *J. Diff. Geom.* **36** (1992), 765–769.

[KM]   Kollár, J. and Mori, S., *Birational Geometry of Algebraic Varieties*, with collaboration of C. H. Clemens and A. Corti, Cambridge Tracts in Mathematics **134**, Cambridge University Press, Cambridge, 1998.

[L]   Lazarsfeld, R., *Positivity in Algebraic Geometry*, in preparation.

[LS]   LeBrun, C. and Salamon, S., Strong ridigity of positive quaternion-Kähler manifolds, *Invent. Math.* **118** (1994), 109–132.

[Lu]   Lübke, M., Stability of Einstein–Hermitian vector bundles, *Manuscripta Math.* **42** (1983), 245–257.

[M]   Matsuki, K., *Introduction to Mori's Program*, Universitext, Springer-Verlag, New York, to appear.

[Ma]   Matsumura, H., *Commutative ring theory*, Second edition, Cambridge Studies in Advanced Mathematics, **8**, Cambridge University Press, Cambridge, 1989.

[Mat]   Matsusaka, T., On canonically polarized varieties (II), *Amer. J. Math.* **92** (1970), 283–292.

[Me]   Megyesi, G., Fano threefolds in positive characteristic, *J. Alg. Geom.* **7** (1998), 207–218.

[Mi]   Miyaoka, Y., Relative deformations of morphisms and applications to fibre spaces, *Comm. Math. Univ. Sancti Pauli* **42** (1993), 1–7.

[MM]   Miyaoka, Y. and Mori, S., A numerical criterion for uniruledness, *Ann. of Math.* **124** (1986), 65–69.

[MP]   Miyaoka, Y. and Peternell, T., *Geometry of Higher Dimensional Varieties*, DMV-Seminar, **26**, Birkhäuser Verlag, Basel, 1997.

[Mo1]   Mori, S., Projective manifolds with ample tangent bundles, *Ann. of Math.* **110** (1979), 593–606.

[Mo2]   Mori, S., Threefolds whose canonical bundles are not numerically effective, *Ann. of Math.* **116** (1982), 133–176.

[MM]   Mori, S. and Mukai, S., Classification of Fano threefolds with $b_2 \geq 2$, *Manuscr. Math.* **36** (1981), 147–162.

[Mu1]   Mumford, D., *Abelian Varieties*, Oxford University Press, London, 1974.

[Mu2]   Mumford, D., *Lectures on Curves on Algebraic Surfaces*, Princeton University Press, Princeton, 1966.

[MuK]   Mumford, D., Varieties defined by quadratic equations, with an appendix by G. Kempf, *Questions on Algebraic Varieties (C.I.M.E., III Ciclo, Varenna, 1969)*, Edizioni Cremonese, Rome.

[N]   Nadel, A., The boundedness of degree of Fano varieties with Picard number 1, *J. Amer. Math. Soc.* **4** (1991), 681–692.

[Na]   Nakayama, N., The lower semi-continuity of the plurigenera of complex varieties, *Algebraic Geometry, Proc. Symp., Sendai 1985*, T. Oda, ed., Adv. Stud. Pure Math. **10**, North-Holland, Amsterdam, 1987.

[P]   Peternell, T., Minimal varieties with trivial canonical classes, I, *Math. Zeit.* **217** (1994), 377–405.

[RC]   Ran, Z. and Clemens, H., A new method in Fano geometry, *Internat. Math. Res. Notices* **10** (2000), 527–549.

[Re]   Reid, M., Canonical 3-folds, *Journées de Géométrie Algébrique d'Angers, Juillet 1979*, A. Beauville, ed., Sijthoff & Noordhoff, Alphen aan den Rijn–Germantown, Md., 1980.

[SB]   Shepherd-Barron, N., Fano threefolds in positive characteristic, *Comp. Math.* **105** (1997), 237–265.

[SK]   Shioda, T. and Katsura, T., On Fermat varieties, *Tôhoku Math. J.* **31** (1979), 97–115.

[Sh1]   Shokurov, V. V., The nonvanishing theorem (in Russian), *Izv. Akad. Nauk SSSR Ser. Mat.* **49** (1985) 635–651; English transl.: *Math. USSR Izvestiya* **19** (1985), 591–604.

[Sh2]   Shokurov, V. V., The existence of a line on Fano varieties (in Russian), *Izv. Akad. Nauk SSSR Ser. Mat.* **43** (1979) 922–964; English transl.: *Math. USSR Izvestiya* **14** (1980), 395–405.

[Si]   Simpson, C., Subspaces of moduli spaces of rank one local systems, *Ann. Sci. École Norm. Sup.* **26** (1993), 361–401.

[U1]   Ueno, K., *Classification Theory of Algebraic Varieties and Compact Complex Spaces*, Lecture Notes in Mathematics **439**, Springer-Verlag, Berlin, 1975.

[U2]   Ueno, K., Geometry of algebraic and analytic threefolds, *Algebraic Threefolds, Proc. of the Second Session of C.I.M.E., Varenna*, A. Conte, ed., Lecture Notes in Mathematics **947**, Springer-Verlag, Berlin, 1982.

[T]   Tsuji, H., Global generation of adjoint bundles, *Nagoya. Math. J.* **142** (1996), 5–16.

[V]   Viehweg, E., Vanishing theorems, *J. für die reine und angewandte Math.* **335** (1982), 1–8.

[W]   Wiśniewski, J., On contraction of extremal rays of Fano manifolds, *J. für die reine und angewandte Math.* **417** (1991), 141–157.

[Wl]   Włodarczyk, J., Toroidal varieties and the weak factorization theorem, preprint.

[Y1]   Yau, S. T., On Calabi's conjecture and some new results in algebraic geometry, *Proc. Nat. Acad. Sci. USA* **74** (1977), 1798–1799.

[Y2]   Yau, S. T., On the Ricci curvature of a compact Kähler manifold and the complex Monge–Ampère equation, I, *Comm. Pure and Appl. Math.* **31** (1978), 339–411.

[Z]   Zhang, Q., On projective manifolds with nef anticanonical bundle, *J. für die reine und angewandte Math.* **478** (1996), 57–60.

Index

Universitext   *(continued)*